## BLACK&DECKER®

### THE COMPLETE GUIDE TO

# GARDEN WALLS &
# FENCES

- **Improve Backyard Environments**
- **Enhance Privacy & Enjoyment**
- **Define Space & Borders**

D1337566

Creative Publishing
international

MINNEAPOLIS, MINNESOTA
www.creativepub.com

**Creative Publishing international**

Copyright © 2010
Creative Publishing international, Inc.
400 First Avenue North, Suite 300
Minneapolis, Minnesota 55401
1-800-328-0590
www.creativepub.com
All rights reserved

Printed in China

10 9 8 7 6 5 4 3 2 1

Library of Congress Cataloging-in-Publication Data

The complete guide to garden walls & fences : improve backyard
environments, enhance privacy & enjoyment, define space &
borders.
    p. cm.
  "Created by the editors of Creative Publishing International, Inc
incooperation with Black & Decker."
  Includes index.
  ISBN-13: 978-1-58923-519-9 (soft cover)
  ISBN-10: 1-58923-519-3 (soft cover)
  1.  Fences. 2.  Garden structures. 3.  Retaining walls. 4.  Walls.  I.
Creative Publishing International. II. Black & Decker Corporation
(Towson, Md.)
  TH4965.C65 2010
  690'.89--dc22
                              2009048792

*President/CEO:* Ken Fund
*VP for Sales & Marketing:* Kevin Hamric

**Home Improvement Group**

*Publisher:* Bryan Trandem
*Managing Editor:* Tracy Stanley
*Senior Editor:* Mark Johanson
*Editor:* Jennifer Gehlhar

*Creative Director:* Michele Lanci-Altomare
*Art Direction/Design:* Jon Simpson, Brad Springer, James Kegley

*Lead Photographer:* Joel Schnell
*Set Builder:* James Parmeter
*Production Managers:* Laura Hokkanen, Linda Halls

*Page Layout Artist:* Danielle Smith
*Shop Help:* Charles Boldt

*Edition Editor:* Phillip Schmidt
*Copy Editor:* Betsy Matheson
*Proofreader:* Kristen Olson

*The Complete Guide to Garden Walls & Fences*
*Created by:* The Editors of Creative Publishing international, Inc., in cooperation with Black & Decker.
Black & Decker® is a trademark of The Black & Decker Corporation and is used under license.

## NOTICE TO READERS

For safety, use caution, care, and good judgment when following the procedures described in this book. The Publisher and Black & Decker cannot assume responsibility for any damage to property or injury to persons as a result of misuse of the information provided.

The techniques shown in this book are general techniques for various applications. In some instances, additional techniques not shown in this book may be required. Always follow manufacturers' instructions included with products, since deviating from the directions may void warranties. The projects in this book vary widely as to skill levels required: some may not be appropriate for all do-it-yourselfers, and some may require professional help.

Consult your local Building Department for information on building permits, codes and other laws as they apply to your project.

# Contents

## The Complete Guide to Garden Walls & Fences

**Introduction** . . . . . . . . . . . . . . . . . . . . . . . . . .5
**Gallery of Garden Walls & Fences** . . . . . . . . . . . . . . . . . .6

**PLANNING & BASIC TECHNIQUES** . . . . . . . . . . . . . . . . . .**15**
Making a Plan. . . . . . . . . . . . . . . . . . . . . . . . . . .16
Handling Slope . . . . . . . . . . . . . . . . . . . . . . . . . .18
Laying Out Fencelines . . . . . . . . . . . . . . . . . . . .22
Setting Posts . . . . . . . . . . . . . . . . . . . . . . . . . .26
Working with Stone . . . . . . . . . . . . . . . . . . . . .30
Working with Brick & Concrete Block . . . . . . . . .34
Mixing & Placing Mortar. . . . . . . . . . . . . . . . . .38
Working with Concrete . . . . . . . . . . . . . . . . . . .40
Concrete Footings. . . . . . . . . . . . . . . . . . . . . . .42
Fence & Wall Materials. . . . . . . . . . . . . . . . . . .46
Tools . . . . . . . . . . . . . . . . . . . . . . . . . . . . . . . .48

**FENCE & GATE PROJECTS** . . . . . . . . . . . . . . . . . . . . . . . .**51**
Board & Stringer Fence . . . . . . . . . . . . . . . . . . .52
Wood Panel Fences . . . . . . . . . . . . . . . . . . . . . .56
Picket Fence. . . . . . . . . . . . . . . . . . . . . . . . . . .64
Post & Board Fences . . . . . . . . . . . . . . . . . . . . .68
Split Rail Fence . . . . . . . . . . . . . . . . . . . . . . . .74
Virginia Rail Fence . . . . . . . . . . . . . . . . . . . . . .78
Wood Composite Fence. . . . . . . . . . . . . . . . . . .82
Vinyl Panel Fence . . . . . . . . . . . . . . . . . . . . . . .86
Ornamental Metal Fence . . . . . . . . . . . . . . . . . .90
Chain Link Fence & Gate . . . . . . . . . . . . . . . . . .94
Trellis Fence . . . . . . . . . . . . . . . . . . . . . . . . . .102
Bamboo Fence. . . . . . . . . . . . . . . . . . . . . . . . .106
Invisible Dog Fence . . . . . . . . . . . . . . . . . . . . .110
Brick & Cedar Fence . . . . . . . . . . . . . . . . . . . .116
Stone & Rail Fence. . . . . . . . . . . . . . . . . . . . . .120
Easy Custom Gates . . . . . . . . . . . . . . . . . . . . .122
Arched Gate . . . . . . . . . . . . . . . . . . . . . . . . . .126
Trellis Gate . . . . . . . . . . . . . . . . . . . . . . . . . . .130

**GARDEN WALLS** . . . . . . . . . . . . . . . . . . . . . . . . . . . . .**135**
Patio Wall . . . . . . . . . . . . . . . . . . . . . . . . . . . .136
Outdoor Kitchen Walls
& Countertop . . . . . . . . . . . . . . . . . . . . . . . . .142
Dry Stone Wall . . . . . . . . . . . . . . . . . . . . . . . .146
Mortared Stone Wall . . . . . . . . . . . . . . . . . . . .150
Brick Garden Wall. . . . . . . . . . . . . . . . . . . . . . .154
Mortarless Block Wall. . . . . . . . . . . . . . . . . . . .158
Poured Concrete Wall . . . . . . . . . . . . . . . . . . .162
Interlocking Block Retaining Wall. . . . . . . . . . . .168
Timber Retaining Wall . . . . . . . . . . . . . . . . . . .174
Stone Retaining Wall . . . . . . . . . . . . . . . . . . . .176
Poured Concrete Retaining Wall . . . . . . . . . . . .180

**REPAIRS FOR WALLS & FENCES**. . . . . . . . . . . . . . . . . .**187**
Stone Walls. . . . . . . . . . . . . . . . . . . . . . . . . . .188
Brick Structures . . . . . . . . . . . . . . . . . . . . . . . .192
Wood Fences . . . . . . . . . . . . . . . . . . . . . . . . . .196

**Resources**. . . . . . . . . . . . . . . . . . . . . . . . . . .**204**
**Photo Credits**. . . . . . . . . . . . . . . . . . . . . . . . .**205**
**Index** . . . . . . . . . . . . . . . . . . . . . . . . . . . . . . .**206**

# Introduction

Stroll through any residential neighborhood and you're bound to see at least a dozen different styles of fencing and garden walls built with nearly as many different materials. If you pause to admire the look of a fence or wall, consider how it enhances (or detracts from) the house and grounds. What do the homeowners achieve by adding the fence or wall to the yard? Perhaps the fence adds privacy for a backyard patio or security for a swimming pool or play area. Or, maybe it creates a decorative boundary line, adding curb appeal.

As a passerby, what does the fence or wall do for you? Does it make you feel welcome or just the opposite? Does it direct visitors to a front entrance? Does it complement the house in the background or simply obscure it from view? These are the kinds of questions you'll be asking yourself when it's time to brainstorm about your new fence or wall project. In other words, there's more to the planning process than choosing a style and materials.

For starters, it's helpful to think of walls and fences as more than just borders and barriers; indeed, they can be powerful design tools. To appreciate what a wall or fence can do for your outdoor spaces, think for a moment about the many uses of walls inside your home. Walls are the primary building blocks of every floor plan. They create rooms out of empty space. They make private areas private and turn wide-open spaces into more usable, comfortable zones. They direct traffic through the house and serve as backdrops for furniture and decorations. And finally, with the help of windows, walls shape and enhance your views of the outside world.

Outdoor walls and fences can have the same transformative effects. A tall, solid fence instantly provides privacy and a sense of security and enclosure, creating a private haven within its borders. An ornamental metal fence adds just as much security without a visual barrier. Both fences and low walls can direct traffic—from the street to the front door, around the house to the backyard, or into and around a garden. Walls make wonderful backdrops for all kinds of ornamental plants, and they're great for defining patio spaces and other seating or entertaining areas. And with just the right amount of open space, a wall, fence, or gate can offer tantalizing glimpses of what lies beyond, adding a sense of mystery to a colorful flower garden or a stunning view in the distance.

Whatever look and functions you have in mind, this book will help you choose the right materials and products and show you how to tackle the job from start to finish. You'll learn traditional building techniques, like bricklaying and dry-stacking wall stones, as well as installation steps for the newest prefab fencing products, including wood composite and ornamental metal fences. Your options are wide open—keep that in mind the next time you're out on an "inspirational" stroll through the neighborhood.

# Gallery of Garden Walls & Fences

Getting just what you want in the finished product of your new fence or wall starts with the easiest step of all: exploring your options. With so many materials, sizes, and design features to choose from, it's wise to spend some time discovering what's out there before you make any big decisions.

The beautiful fences and walls shown here are just a small sample of some things you might do with your own project. Other places to gather ideas include books and magazines on garden and landscape design, websites of fencing and wall materials manufacturers, and even stock photography websites (using search terms such as "garden walls," "fence," "picket fence," etc.).

As mentioned before, the single best way to generate ideas for your project is by touring some neighborhoods in your area. There, you'll find a living gallery of fences and walls of all descriptions. Be sure to examine not only the structures themselves, but also how each fits into the context of its host property—good examples can inspire you with new ideas and insights, while bad examples might serve as a cautionary message about what not to do.

**Long-prized for its strength,** combined with a fineness of detail and gorgeous patterning, wrought iron fencing remains a symbol of craftsmanship and elegance. New iron fencing is still available today and is sold through specialty retailers and fabricators.

**Like the background** in a landscape painting, fences and walls provide context and structure for the surrounding features. The interplay between the wildness of plants and the architectural lines of a fence or wall can be stunning.

**Utilitarian by design, yet pleasing to the eye,** the post and board fence evokes the uncomplicated beauty and peacefulness of rolling countryside. The same effect holds true in suburban settings.

**Much more stylish than the sum of its parts,** this custom privacy fence is made with standard sizes of cedar lumber, using basic post and board construction. The posts are capped with vertical cedar boards to conceal the siding joints and create a trim, finished look overall.

**Landscape block**—cast concrete units that can look like stone, adobe, or brick—is well on its way to becoming the next great outdoor building material. It's extremely durable and easy to build with and is much less expensive than natural stone.

**Simple embellishments** can add a great deal of character to a standard wood privacy fence. You can easily trim the tops of your siding boards after the fence is up, using a jigsaw. Here, the fence posts were given prominence to punctuate the rhythm of the wave effect and provide a structural element for the composition.

**Covering walls in gardens and outdoor rooms the world over, stucco** offers a pleasing combination of a finished look with an organic texture and feel. Behind the stucco finish of most outdoor walls is a simple, durable structure of mortared or unmortared concrete block.

**The gate is a prime opportunity** for adding character and charm. This beautiful arbor gate not only makes for a grand portal, it adds another dimension to the fence and provides a welcome contrast to the bold pattern of the lattice panels.

**The right fence material makes all the difference.** Ornamental metal's clean lines and permanence are the perfect complement to this home's spare detailing and formal character. Using similar metal materials for the balcony, handrail, and fence unifies the look of the whole property.

**Bamboo's exceptional strength and light weight** make it suitable for infill on gates and other fence features that take a beating. This bamboo gate was built with the same structure and materials as the fence for a nicely integrated look.

**Poured concrete** might not be the first material that comes to mind when dreaming of a new garden wall, but it's certainly worth consideration. The versatility of concrete can inspire all sorts of custom creations, such as this retaining wall with a traditional frame-and-panel effect.

**A mortared block retaining wall** provides for easy passage along the edge of this lavishly planted property. Opposite the wall, a split-cedar post and rail fence creates a clear border as it complements the view of the wooded area beyond.

**A wall or fence surrounding a home's entry door** instantly creates an entrance courtyard or semi-private patio space. Depending on the height of the wall and presentation of the gate, the barrier can be welcoming to visitors or send the message that the space within is private.

**A fence that's always current:** as an alternative to wood planks, the builder of this attractive fence used electrical conduit (a.k.a. EMT, for electrical metal tubing) as infill. The tubes are held by predrilled 2 × 2s set inside the 2 × 4 stringers. Rigid copper pipe is another good material for this kind of fence.

**The classic picket fence** is the perfect example of how a low, decorative fence or wall separates a property from the outside world while at the same time it enhances the view of all that lies within its boundary.

**Fences make their greatest design impact** by defining and influencing our perception of space. Even an unassuming fence like this can have several effects—dividing the public and private zones of the property, guiding visitors to the entrance point, and highlighting the flowers at its base.

**Some settings call for a break** with conventional fence design and materials. Here, tropical hardwood siding boards were rounded over along the edges and finished with a protective coating to highlight the wood's natural coloring. Painting the fence posts to match the house offsets the dark wood tones and lends a custom-built look to the fence.

**When it comes to retaining walls,** interlocking concrete block is the ultimate in user-friendly. Just stack it up to create walls up to 3 ft. high in any shape or configuration you need. Tame tall slopes with multiple low walls, adding terraced garden spaces in the process.

**Whether it's gathered right from the surrounding ground** or shipped to you on a pallet, stone is nature's ideal building block for garden creations. A dry-stacked wall like this is a great do-it-yourself project with a truly timeless history.

**Vinyl fences** today are usually guaranteed for decades not to rot, warp, or discolor. They are made with a plastic compound.

# Planning
# & Basic
# Techniques

Walls and fences are great do-it-yourself projects, but they often require a little more planning than other outdoor improvements. Perhaps most importantly, you'll need to make sure your new structure will stand within your property lines. Careful planning is also important for more personal reasons, such as making sure the finished product meets your needs while adding just the right touch to your house and landscape.

When it comes to the design and layout phase of the project, it's a good idea to work things out on paper, when mistakes and miscalculations can still be corrected with an eraser and you're free to run with new ideas to see where they take you. Mapping out the structure with dimensions is also invaluable for working up an accurate materials list and estimating costs. The total cost of fencing materials, in particular, can be a big factor when choosing a fence style and material.

In addition to planning and materials information, this chapter covers many of the basic construction techniques essential to completing the projects in this book. You'll also find tips for working with specialty materials, like stone, brick, and poured concrete.

## In this chapter:

- Making a Plan
- Handling Slope
- Laying Out Fencelines
- Setting Posts
- Working with Stone
- Working with Brick & Concrete Block
- Mixing & Placing Mortar
- Working with Concrete
- Concrete Footings
- Fence & Wall Materials
- Tools

# Making a Plan

**Take overall measurements of your yard** and measure distances between objects in the yard before you begin to draw your site plan.

Making a plan begins with taking measurements in your yard. With accurate measurements you can draw a detailed scale drawing of your yard, called a site plan. Note the boundaries of your property on the plan, along with permanent structures, sidewalks, trees, shrubs and planting beds. You should focus on the areas where you'll be building, but it's important to get the whole yard into the plan. By going through this process, you may get a new perspective on how your project fits in (or doesn't fit in) to your current landscape.

You should also physically mark your property lines as you take measurements. If you don't have a plot drawing (available from the architect, developer, contractor, or the previous homeowner) or a deed map (available from city hall, county courthouse, title company, or mortgage bank) that specifies property lines, hire a surveyor to locate and mark them. File a copy of the survey with the county as insurance against possible boundary disputes in the future.

## Building Codes and Utilities ▶

Before you can even begin drawing plans for your fence, wall, or gate, you need to research local building codes. Building codes will tell you if a building permit and inspection are needed for a project. Some code requirements are designed to protect public safety, while others help preserve aesthetic standards.

Codes may dictate what materials can be used, maximum heights for structures, depths for concrete footings and posts, and setback distance or how far back a fence or wall must be from property lines, streets, or sidewalks. Setback distance is usually 6 to 12 inches and is especially important on a corner lot, since a structure could create a blind corner. A fence or wall may be built directly above a property line if agreed by both neighbors who share ownership of the fence.

If you find a fence, wall, or gate design that appeals to you, but does not meet local ordinances, the municipal authorities may be willing to grant a variance, which allows you to compromise the strict requirements of the code. This normally involves a formal appeal process and perhaps a public hearing.

Another thing to consider as you plan your project is the placement of any utility lines that cross your property. At no cost, utility companies will mark the exact locations and

depths of buried lines so you can avoid costly and potentially life-threatening mistakes. In many areas, the law requires that you have this done before digging any holes. Even if not required by law in your area, this step is truly necessary.

A fence, wall, or gate on or near a property line is as much a part of your neighbors' landscapes as your own. As a simple courtesy, notify your neighbors of your plans and even show them sketches; this will help to avoid strained relationships or legal disputes. You may even decide to share labor and expenses, combining resources for the full project or on key features that benefit you both.

**Consult your electric utility office,** phone company, gas and water department, and cable television vendor for the exact locations of underground utility lines.

# Drawing the Plan

Good plans make it possible to complete a project efficiently. Plotting fence, wall, and gate locations on paper makes it much easier to determine a realistic budget, make a materials list, and develop a realistic work schedule. *Tip: Don't start drawing onto your site plan right away. Make a number of photocopies that you can draw on, and save the original as a master copy.*

Along with your site map, an elevation chart may be helpful if you have significant slope to contend with. On a copy of the site map, locate and draw the fence or wall layout. Consider how to handle obstacles like large rocks and trees or slopes. Be sure you take into account local setback regulations and other pertinent building codes.

As you begin to plot your new fence or wall, you'll need to do a little math. To determine the proper on-center spacing for fence posts, for example, you divide the length of the fence into equal intervals—six to eight foot spacing is typical. If your calculations produce a remainder, don't put it into one odd-sized bay. Instead, distribute the remainder equally among all the bays or between the first and last bay (unless you are installing prefab panels).

If you plan to use prefabricated fence panels, post spacing becomes more critical. If you'd like to install all your posts at once (the most efficient strategy), you'll need to add the width of the post to the length of the panel plus an extra half-inch for wiggle room in your plan. But most fence panel manufacturers suggest that you add fence posts as you go so you can locate them exactly where the panels dictate they need to be.

If you're making a plan for building a wall, be sure to plan enough space around the wall itself for footings that are at least twice as wide as the wall they will support. Carefully plot each corner and curve, and allow plenty of space between the footings and obstacles such as trees or low-lying areas where water may collect.

Once you've worked out the details and decided on a final layout, convert the scale dimensions from the site map to actual measurements. From this information, draw up a materials estimate, adding 10 percent to compensate for errors and oversights.

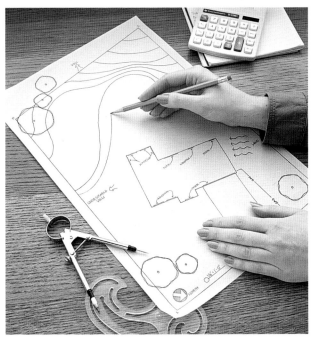

**A site map** is an overhead view of a fence, wall, or gate setting drawn to scale. It aids in the visualization and planning of a project. From the measurements of your yard survey, convert all actual measurements to scale measurements (if ⅛" = 1 ft., multiply actual measurements by .125). On paper, draw straight boundaries to scale. Scribe arcs with a compass to mark triangulated points, noting the edges and corners of permanent structures, such as your house or garage. Use these points as established references to plot all the elements in the property. To finish the site map, draw contour lines to indicate slope, and mark compass directions, wind patterns, utilities, and any other pertinent information that will influence the location of your fence, wall, or gate.

**Make an elevation drawing** of your yard if it has a significant slope. Measure the vertical drop of a slope using stakes and string. Connect the string to the stakes so it is perfectly horizontal. Measure the distance between the string and ground at 2 ft. intervals along the string.

# Handling Slope

It's considerably easier to build a fence or garden wall when the ground is flat and level along the entire length of the proposed site line. But few landscapes are entirely flat. Hills, slight valleys, or consistent downward grades are slope issues to resolve while planning your fence. There are two common ways to handle slope: contouring and stepping.

With a contoured fence, the stringers are parallel to the ground, while the posts and siding are plumb to the earth. The top of the fence maintains a consistent height above grade, following the contours of the land. Most pre-assembled panel fences cannot be contoured, since the vertical siding members are set square to the stringers. Picket fence panels may be "racked" out of square for gentle contouring. Vinyl fence sections generally permit contouring.

Each section of a stepped fence is level across the top, forming the characteristic steps as the ground rises or falls. Stepped fences appear more structured and formal. Pre-assembled panels may be stepped to the degree their bottoms can be trimmed for the slope, or that additional siding (such as kick boards) can be added to conceal gaps at the tall end of the step. Stepped custom-built fences are more work than contoured fences since vertical siding boards must be trimmed to length individually and post heights may vary within a layout.

**Solid planning and careful execution** allow you to turn a sloped yard into a positive design factor when you build your fence or wall project.

# Strategies for Managing Slope

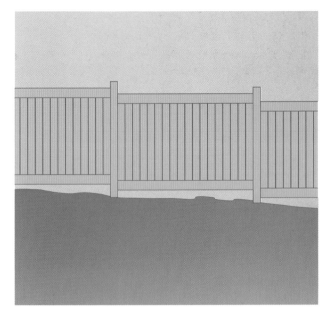

**Stepped panels** are horizontal, maintaining an even height between posts. A good strategy for pre-built panel systems, stepping fences is the only way to handle slope when working with panels that cannot be trimmed, racked, or otherwise altered.

**Racking a panel** involves manipulating a simple fence panel by twisting it out of square so the stringers follow a low slope while the siding remains vertical. Stockade and picket panels are good candidates for this trick, but the degree to which you can rack the panels is limited. If the siding is connected to stringers with more than one fastener at each joint, you'll need to remove some fasteners and replace them after racking the panel.

**Contouring** creates a more casual, natural-looking fence. Each individual siding board is set the same distance from the ground below and allowed to extend to full height without trimming. The resulting top of the fence will mimic the ground contour.

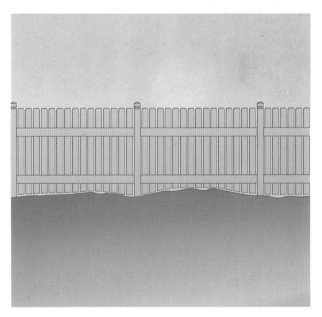

**Bottom trimming** creates a level fenceline with a baseline that follows the slope and contour of the land. On low slopes you can use this technique and trim the siding boards on pre-made panels that have open bottoms (in some cases you can raise the bottom stringer). Bottom trimming is best for site-built board and stringer fences, however.

# Contoured Fences

A contoured fence rolls along with the terrain, maintaining a consistent height above the ground as it follows the land. Picket fences and others with individual siding work best for contouring. There are multiple tactics you can use to build a contoured fence. The scenario described below involves setting all your posts, installing stringers, trimming posts to uniform height above the top stringer and then adding the picket siding.

## CONTOURED INSTALLATION OVERVIEW

Begin the layout by running a string between batter boards or stakes located at the ends and corners of the fenceline, adding intermediate batter boards or stakes as needed to keep the string roughly parallel with the grade. Mark the post centers at regular distances (usually six or eight feet) on the string. Don't forget to allow for the posts when measuring. Drop a plumb bob at each post mark on the string to determine posthole locations. Mark these locations with a piece of plastic pegged to the ground or by another method of your choosing.

Align, space, and set the posts (if appropriate for your fence type). Attach the lower stringers between posts. If you are using metal fence rail brackets, bend the lower tab on each bracket to match the slope of the stringer. Each stringer should follow the slope of the ground below as closely as possible while maintaining a minimum distance between the highest point of the ground and the bottom of the stringer. This distance will vary from fence to fence according to your design, but 12 inches is a good general rule.

Install all of the lower stringers and then install the upper stringers parallel to the lower ones. Make sure to maintain an even spacing between the stringers. Establish the distance from the upper stringer to the post tops and then measure this distance on each post. Draw cutting lines and trim the post tops using a circular saw and a speed square clamped to the post as a guide.

Make a spacer that's about the same width as the siding boards, with a height that matches the planned distance from the ground to the bottom of each siding board. Set the spacer beneath each board as you install it. You'll also want a spacer to set the gap between siding boards. Install the siding and add post caps.

**Slope Option 1: Contour**

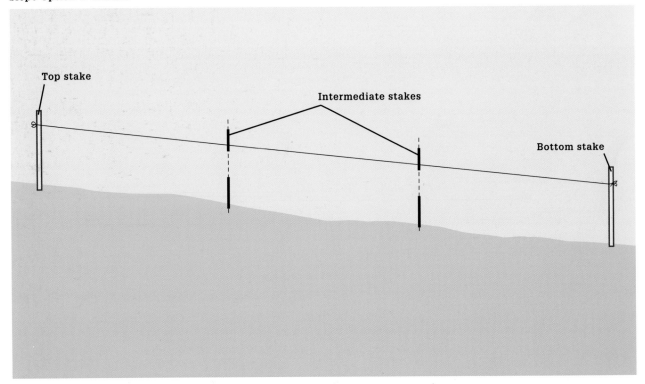

**Contoured fences** can follow ground with either a regular slope or an irregular slope. Place a stake at the beginning and end of the fenceline and at each corner. Add intermediate stakes to maintain spacing when the slope changes.

# Stepped Fences

A stepped fence retains its shape and configuration regardless of changes in slope. The effect of the stepping up or down of whole panels it to create a more formal appearance, but it also lets you avoid cutting premade fence panels. The sacrifice is that you often end up with very tall fence posts and you may need to add filler wood between panel bottoms and irregular dips in the ground.

The following stepping technique works over slopes of a consistent grade. If the grade changes much, bracket each new slope with its own stake or pair of batter boards, as in the illustration on the previous page. Treat the last post of the first run as the first post of the second run and so on.

Alternatively, step each section independently, trimming the post tops after the siding is set. The scenario below describes a flat cap stringer, which some fences use to create a smooth top. If this is not needed on your fence, simply measure down the appropriate distances to position the inset or face-mounted stringers.

## STEPPED INSTALLATION OVERVIEW

Using mason's string and stakes or batter boards, establish a level line that follows the fenceline. Measure the length of the string from end stake to end stake. This number is the run. Divide the run into equal segments that are between 72 and 96". This will give you the number of sections and posts (number of sections plus one).

Measure from the ground to the string at both end stakes. The difference between the two measurements is the rise of the slope. Divide the rise by the number of fence sections on the slope to find the stepping measurement.

Measure and mark the post locations along the level string with permanent marker "Vs" on tape. Drop a plumb bob from each post location mark on the string. Mark the ground with a nail and a piece of bright plastic.

Set the first post at one end and the next one in line. Mark the trim line for cutting to height and run a level string from the cutting line to the next post. Measure up (or down) from the string for the step size distance. Adjust marks if necessary before cutting the posts.

Repeat until you reach the end of the fenceline. Avoid creating sections that will be too tall or too short. The bottom stringer should remain at least four inches above grade.

Cut all posts and then attach stringers or panels so the distance from the tops of the posts to the stringers is consistent.

**Slope Option 2: Stepping**

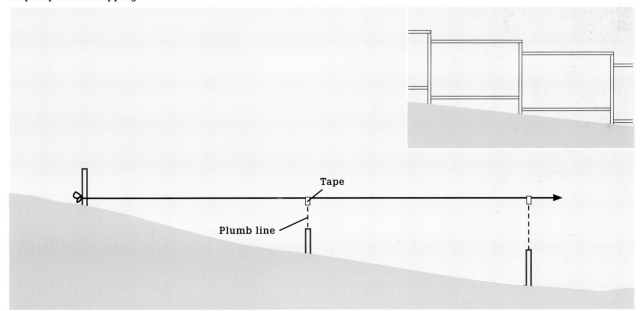

**Stepped fences** (inset) can be installed on either regular or irregular slopes. To plan the fence, run a mason's string between stakes or batter boards at the high end and the low end of the fenceline; measure the distance from the string to the ground at both ends, then calculate the difference between measurements to find the total rise. Divide this amount by the number of fence sections to determine the stepping measurement for each fence panel. On irregular slopes, the amount of drop will vary from section to section

# Laying Out Fencelines

Fence installations begin with plotting the fenceline and marking post locations. Make a site map and carefully measure each post location. The more exact the posthole positions, the less likely it is that you'll need to cut stringers and siding to special sizes.

For walls, determine the outside edges of the footings along the entire site, as for a fenceline. Then plot right angles to find the ends and inside edges of the footings.

Laying out a fence or wall with square corners or curves involves a little more work than for a straight fenceline. The key for these techniques is the same as for plotting a straight fenceline: measure and mark accurately. This will ensure proper spacing between the posts and accurate dimensions for footings, which will provide strength and support for each structure.

## Tools & Materials ▶

Stakes & mason's string
Line level
Tape measure (2)
Level
Circular saw
Hammer
Masking tape
Work gloves

Eye and ear protection
Pencil
Spray paint
Hand maul
1 × 4, 2 × 4 lumber
Permanent marker
Screw gun
Screws

**Use a pair of wood stakes** and mason's string to plot the rough location of your fence or wall. Then, for greater accuracy, install batter boards to plot the final location.

## How to Lay Out a Straight Fenceline

Determine your exact property lines. Plan your fenceline with the locally required setback (usually 6 to 12 inches) from the property line, unless you and your neighbor have come to another agreement. Draw a site map. It should take all aspects of your landscape into consideration, with the location of each post marked. Referring to the site map, mark the fenceline with stakes at each end or corner-post location.

Drive a pair of wood stakes a couple of feet beyond each corner or end stake. Screw a level crossboard across the stakes about six inches up from the ground on the highest end of the fence run. Draw a mason's string from the first batter board down the fenceline. Level the line with a line level and mark the height of the line against one stake of the second batter board pair. Attach a level batter board to these stakes at this height and tie the string to the crossboard so it is taut.

To mark gates, first find the on-center spacing for the gateposts. Combine the width of the gate, the clearance necessary for the hinges and latch hardware, and the actual width of one post. Mark the string with a "V" of masking tape to indicate the center point of each gatepost.

To mark remaining posts, refer to your site map, and then measure and mark the line post locations on the string with marks on masking tape. Remember that the marks indicate the center of the posts, not the edges.

# How to Install Batter Boards

**1** First pair of stakes

Crossboard

**2** Second pair of stakes

Line from first pair of stakes

**3** Post location

**To install batter boards,** drive a pair of short wood stakes a couple of feet beyond each corner or end of the rough planned fenceline. Screw a level crossboard across one pair of stakes, about 6" up from the ground on the higher end of the fence run. Loosely tie a mason's string to the middle of the crossboard.

**Stretch the mason's line** from the batter board to the second pair of stakes at the opposite end or corner of the run. Draw the string tight, and level it with a line level. Mark the string's position onto one of the stakes. Fasten a crossboard to the second pair of stakes so it is level and its top is aligned with the mark on the stake. Tie the mason's line to the center of the crossboard.

**Measure out** from the starting points of the fenceline and mark post locations directly onto the layout lines using pieces of masking tape (don't forget to allow for the widths of your posts—see tips below).

## Tips for Spacing Line Posts and Gate Posts ▸

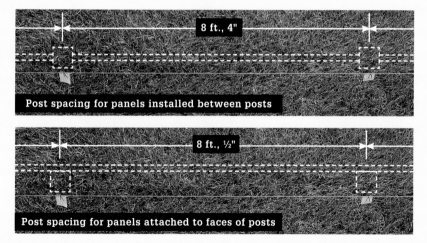

8 ft., 4"

Post spacing for panels installed between posts

8 ft., ½"

Post spacing for panels attached to faces of posts

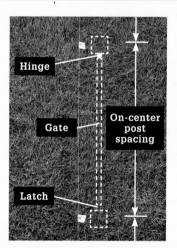

Hinge

Gate

On-center post spacing

Latch

**If your fence panels will be installed between fence posts** (top photo) and you are using 4 × 4 wood posts, add 4" to the length of the fence panels and use that distance as the on-center span between posts (the 4 × 4 posts are actually only 3½" wide but the extra ½" created by using the full 4" dimension will create just the right amount of "wiggle room" for the panel). If panels will be attached to the post faces, add ½" to the actual panel width to determine post spacing.

**To find the on-center spacing** of gate posts, add the gate width, the clearance needed for hinge and gate hardware, and the actual diameter of one post.

# Laying Out Right Angles

If your fence or wall will enclose a square or rectangular area, or if it joins a building, you probably want the corners to form 90 degree angles. There are many techniques for establishing a right angle when laying out an outdoor project, but the 3-4-5-triangle method is the easiest and most reliable. It is a simple method of squaring your fence layout lines, but if you have the space use a 6-8-10 or 9-12-15 triangle. Whichever dimensions you choose, you'll find it easier to work with two tape measures to create the triangle.

## How to Lay Out a Right Angle

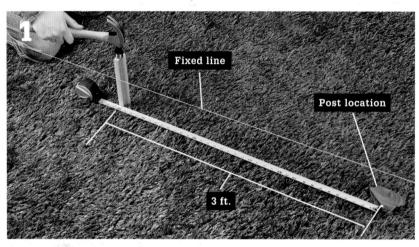

**Drive a pair of stakes along a known fenceline** and run a line that crosses the corner post location (this line should stay fixed as a reference while you square the crossing line to it). Drive a stake 3 ft. out from the corner post location, on the line you don't want to move. You will adjust the other line to establish the right angle.

**Draw one tape measure** from the post location roughly at a right angle to the fixed line. Draw the tape beyond the 4 ft. mark and lock it.

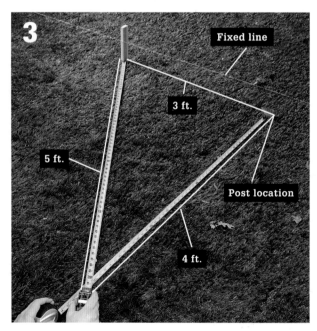

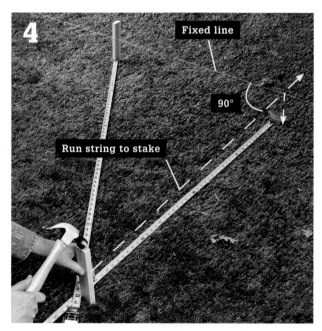

**Angle the second tape measure** from the 3-ft. stake toward the 4 ft. mark on the first tape measure. The two tapes should intersect at 5 ft. and 4 ft.

**Drive a stake** at the point where the tape measure marks intersect. Run a line for this stake to another stake driven past the corner post location to establish perpendicular layout lines. The string tied to the second stake should pass directly over the post location.

# Laying Out Curves

A curve in a fenceline or wall must be laid out evenly for quality results. One easy way to accomplish this is to make a crude compass by tying one end of a string around a can of marking paint and tying the other end to a wood stake, as shown in step 3 below. The radius of the curve should equal the distance from the compass' pivot stake to the starting points of the curve, so make sure to tie the string to this length.

## How to Lay Out a Curve

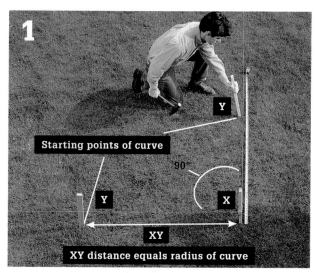

**Plot a right angle at the corner of the outline,** using the 3-4-5 method (see page 24). Measure and drive stakes equidistant from the outside corner to mark the starting points for the curve (labeled "Y" here).

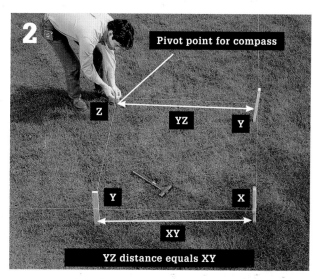

**Tie a mason's string to each Y stake,** and extend the strings back to the corner stake (2). Hold the strings tight at the point where they meet. Then, pull the strings outward at the meeting point until they are taut. Drive a stake at this point to create a perfect square. This stake (labeled "Z" here) will be the pivot point for your string compass.

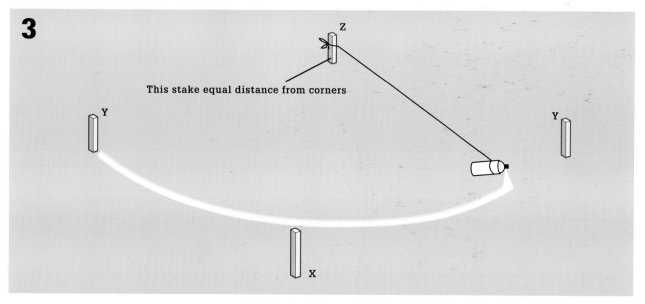

**Mark the curve:** Tie a mason's string to the pivot point (Z) and to a can of marking paint. When the string is held taut, the can's spray nozzle should be even with the stakes at the start of the curve (Y). Keeping the string taut, spray the ground in a smooth arc, extending the curve between the two Y stakes.

# Setting Posts

**Taking the time** to make sure posts are vertical and positioned precisely is perhaps the most important aspect of a successful fence building project.

Even among professional landscapers you'll find widely differing practices for·setting fence posts. Some take the always-overbuild approach and set every post in concrete that extends a foot past the frostline. Others prefer the impermanence, adjustability and drainage of setting posts in packed sand or gravel. Some treat the post ends before setting the posts, others don't bother. The posts may be set all at once, prior to installing the stringers and siding; or, they may be set one at a time in a build-as-you-go approach. Before deciding which approach is best for your situation, it's a good idea to simply walk around your neighborhood and see how the posts for similar fences are installed, then assess which posts seem to be holding up the best.

Another area of dispute is at which point in the process posts should be cut to length (height). While there are those who advocate cutting all posts before installation and then aligning them in the ground before setting them (especially when installing chain link), the most reliable method is to trim the posts to height with a circular saw or handsaw after they are set in the ground and the concrete has set.

Here are some additional thoughts to help you decide how to set your posts:

- Tamped earth and gravel post setting have been increasing the life span and stability of posts for thousands of years by keeping the immediate surroundings of the post drier and firmer.
- The shallow, dish-shaped concrete footing breaks all the rules, but is often the only footing that works in very loose sandy soils. Check with local fence contractors to make sure it's right for your area.
- Hybrid footings help stabilize posts in deep-freezing soils. Quick-set concrete mix may be poured into the hole dry, followed by water (or not, according to local custom—soil moisture is sometimes sufficient to harden the concrete).
- Common posts are set high enough to be trimmed down to their final height. Posts with precut mortises (such as split rail fence posts) or finials need to be set to the final height in the hole.
- Dig holes two times the post thickness for sand-set or gravel-set and closer to three times the diameter if concrete-set.
- For long-term strength, set all gate posts and end posts in concrete.

## Tools & Materials ▶

| | |
|---|---|
| Plumb bob | Masking tape |
| Stakes | Tape measure |
| Hand maul | Speed square |
| Power auger or | Colored plastic |
| posthole digger | Nails |
| Shovel | Eye and ear protection |
| Coarse gravel | Screwgun |
| Carpenter's level | Screws |
| Concrete | Wheelbarrow |
| Mason's trowel | Circular saw |
| 4 × 4 posts | Clamps |
| Scrap lengths of 2 × 4 | Pencil |
| Batter boards & | Work gloves |
| mason's string | Waxed paper |
| Post level | |

The most reliably long-lasting wood posts are pressure-treated with chemicals and labeled for ground contact. Species that are naturally rot resistant are unfortunately less so today than in yesteryear.

Once you've plotted your fenceline with batter boards and string, mark and dig the postholes. Remove the string for digging, but leave the batter boards in place; you will need these for aligning the posts when you set them.

As a general rule, posts should be buried at a minimum depth equal to ⅓ of the total post length (e.g., a post for a six-foot-tall fence will be approximately nine feet long, with three feet buried in the ground). Check with your city's building department for the post depth and burial method required by the local Building Code. Posts set in concrete should always extend below the frost line.

# How to Set Fence Posts

**Set batter boards at both ends of the fenceline.** String a mason's line between the batter boards and level it. Mark post locations on the string with masking tape according to your plan.

**Transfer the marks** from the string to the ground, using a plumb bob to pinpoint the post locations. Pin a piece of colored plastic to the ground with a nail at each post location.

**Dig postholes using a clamshell-type posthole digger** (left photo) or a rented power auger (right photo). Posthole diggers work well for most situations, but if your holes are deeper than 30" you'll need to widen the hole at the top to operate the digger, so consider using a power auger. Make a depth gauge by tacking a board onto a 2 × 4 at the hole depth from the end of the 2 × 4. As you dig, check the depth with the gauge. If you'll be filling the posthole with concrete, widen the bottoms of the holes with your posthole digger to create bell shapes. This is especially important in locales where the ground freezes.

(continued)

**4**

Original position of line

**Reset the mason's string** as a guide for aligning posts. If you want the post to be in exactly the same spot it was laid out, shift the string half the thickness of the post. Pour a 6" layer of gravel into each hole for improved drainage. Position each post in its hole.

## Tubular Forms ▶

**For full concrete footings** in frost-heave-prone soils, cut 8"-dia. concrete forming tubes into 18" sections to collar the posts near grade level and prevent the concrete from spreading. Holes tend to flare at the top, giving concrete footings a lip that freezing ground can push against.

## Mixing Concrete ▶

**If you've never filled postholes** with concrete before, you will be amazed at how much it takes to fill a hole. A 12"-dia. hole that's 36" deep will require around three cubic feet of concrete—or, about six 60-pound bags of dry mix. If you're installing 10 posts, that's 60 bags. This is yet another reason why setting posts one at a time is a good idea—you can spread out the heavy labor of mixing concrete in a wheelbarrow or mortar tub. If you'll be needing more than one cubic yard (27 cubic feet), consider having ready-mix concrete trucked in. But make sure all your posts are braced and set and have at least two wheelbarrows and three workers on hand.

**5**

**Align your post along one line** (or two if it's a corner post). Brace the post on adjacent sides with boards screwed to wood stakes. Adjust to plumb in both directions, anchoring each brace to the post with screws when plumb. As you plumb the post, keep the post flush against the line. Set the remaining posts the same way.

**6**

**Mix concrete in a wheelbarrow** and tamp into the hole with a 2 × 4 to pack the concrete as tightly as you can. Recheck the post alignment and plumb as you go, while correction is still possible. *Tip: Mask the post with waxed paper near the collaring point of the concrete to keep the visible portion of the post clean. Remove the waxed paper before the concrete sets up.*

## Water Runoff ▸

**Form a rounded crown of concrete** with your trowel just above grade to shed water.

**7**

**For reasonably level ground,** draw a mason's string from end post to end post at the height the posts need to be cut (for custom fences, this height might be determined by your shortest post). Mark each post at the string. Carry the line around each post with a pencil and speed square.

**8**

**Wait at least a day for the concrete to set up** and then clamp a cutting guide to the posts (a speed square is perfect). Cut along the trim line on each face of each with a circular saw to trim your posts (this is a great time to use a cordless circular saw). In most cases, you'll want to add a post cap later to cover the end grain.

# Working with Stone

The stone wall projects in this book call for a few basic stoneworking techniques, in addition to the building steps shown each project. The methods of laying stone are as varied as the stone masons who practice the craft. But all of them would agree on a few general principles:

- Thinner joints are stronger joints. Whether you are using mortar or dry-laying stone, the more contact between stones, the more resistance to any one stone dislodging.
- Tie stones are essential in vertical structures, such as walls or pillars. These long stones span at least two-thirds of the width of the structure, tying together the shorter stones around them.

- When working with mortar, most stone masons point their joints deep for aesthetic reasons. The less mortar that is visible, the more the stone itself is emphasized.
- Long vertical joints, or head joints, are weak spots in a wall. Close the vertical joints by overlapping them with stones in the next course, similar to a running bond pattern in a brick or block wall.
- The sides of a stone wall should have an inward slope (called batter) for maximum strength. This is especially important with dry-laid stone. Mortared walls need less batter. See pages 38 to 39 for tips on working with mortar.

See page 49 for a description of the tools you'll need for a successful stoneworking project.

## Estimating Stone Tonnage ▶

To make rough estimates of the amount of stone you'll need for a garden wall project, use the following calculations:

Ashlar stone walls: Area of the wall face (sq. ft.) ÷ 15 = tons of stone needed

Rubble (or irregular) stone walls: Area of the wall face (sq. ft.) ÷ 35 = tons of stone needed

Always add at least 10 percent to your materials estimate when ordering stone. This provides some extra material for practicing cutting and dressing techniques and allows for waste from miscuts and routine trimming.

**Thin joints are the strongest.** When working with mortar, joints should be ½ to 1" thick. Mortar is not intended to create gaps between stones, but to fill the inevitable gaps and strengthen the bonds between stones. Wiggle a stone once it is in place to get it as close as possible to adjoining stones.

# Laying Stone Walls

**Tie stones** are long stones that span most of the width of a wall, tying together the shorter stones and increasing the wall's strength. As a guide, figure that 20 percent of the stones in a structure should be tie stones.

**A shiner is the opposite of a tie stone**—a flat stone on the side of a wall that contributes little in terms of strength. A shiner may be necessary when no other stone will fit in a space. Use shiners as seldom as possible, and use tie stones nearby to compensate.

**Lay stones in horizontal courses,** where possible, a technique called ashlar construction. If necessary, stack two to three thin stones to match the thickness of adjoining stones.

**With irregular stone,** such as untrimmed rubble or field stone, building course by course is difficult. Instead, place stones as needed to fill gaps and to overlap the vertical joints.

**Use a batter gauge** and level to lay up dry stone structures so the sides angle inward. Angle the sides of a wall 1" for every 2 ft. of height—less for ashlar and freestanding walls, twice as much for round stone and retaining walls.

# Cutting Stone

**Build a banker for cutting stone.**
This is a simple sand-bed table that provides a sturdy, shock-absorbent work surface that aids with chisel cuts. To construct a banker, build two square frames out of 2 × 2s and sandwich a matching piece of ¾" plywood between the frames. Fasten the pieces together with 3½" deck screws driven through both sides. Fill one side of the banker with sand to complete the work surface. If desired, set the banker atop a stable base of concrete blocks. *Note: Always wear eye protection when cutting or dressing stone.*

**A circular saw** lets you precut stones with broad surfaces with greater control and accuracy than most people can achieve with a chisel. It's a noisy tool, so wear ear plugs, along with a dust mask and safety goggles. Install a toothless masonry blade on your saw and start out with the blade set to cut ⅛" deep. (Make sure the blade is designed for the material you're cutting. Some masonry blades are designed for hard materials like concrete, marble, and granite. Others are for soft materials, like concrete block, brick, flagstone, and limestone.) Wet the stone before cutting to help control dust, then make three passes, setting the blade ⅛" deeper with each pass. Repeat the process on the other side. A thin piece of wood under the saw protects the saw foot from rough masonry surfaces. Remember: Always use a GFCI-protected outlet or extension cord when using power tools outdoors.

**Laying stones works best** when the sides (including the top and bottom) are roughly square. If a side is sharply skewed, score and split it with a pitching chisel, and chip off smaller peaks with a pointing chisel or mason's hammer. Remember: a stone should sit flat on its bottom or top side without much rocking.

**"Dress" a stone,** using a pointing chisel and maul, to remove jagged edges or undesirable bumps. Position the chisel at a 30 to 45° angle at the base of the piece to be removed. Tap lightly all around the break line, then more forcefully, to chip off the piece. Position the chisel carefully before each blow with the maul.

## How to Cut Flagstone for Walls & Caps

**Mark the stone for cutting** on both sides, using chalk or a crayon. If there is a fissure nearby, mark your line there, since the stone will likely break there naturally. *Note: To prevent unpredicted breaks when cutting off large pieces, plan to chip off small sections at a time.*

**Score along the cut line** on the back side of the stone (the side that won't be exposed) by moving a stone chisel along the line and striking it with moderate blows with a maul. As an alternative, you can do this step with a circular saw.

**Break the stone to complete the cut:** First, turn the stone over and rest it on a metal pipe or a 2 × 4 so the scored edge is directly over the support. Then, strike forcefully near the end of the waste portion to break the stone along the cut line.

# Working with Brick & Concrete Block

Success with any brick or block wall project starts with careful preparation and planning. Most importantly, you want to make sure the wall will have a proper foundation—whether that's a deep concrete footing or a reinforced slab—to prevent cracking and failure due to ground movement, an all-too-common problem with masonry structures.

When it comes to planning and laying out the project, it's always a good idea to complete a dry run by setting down the entire first course of brick or block. This is the most foolproof way to check dimensions, spot potential problems, and mark accurate layout lines for the installation.

All of the masonry projects in this book call for mortar (even the Mortarless Block Wall, which has mortar underneath its first course and within the cavities of its top course). Getting the mortar right is critical to both its strength and workability; the tips on pages 38 and 39 tell you what you need to know.

**Select a construction design** that makes sense for your project. There are two basic methods used in stacking brick or block. Structures that are only one unit wide are called single wythe, and are typically used for projects like brick pillars or planters, and for brick veneers. Double-wythe walls are two units wide and are used in free-standing applications. Most concrete-block structures are single wythe.

## Planning Brick & Block Projects

**Keep structures as low as you can.** Local codes require frost footings and additional reinforcement for permanent walls or structures that exceed maximum height restrictions. You can often simplify your project by designing walls that are below the maximum height.

**Add a lattice panel or another decorative element** to permanent walls to create greater privacy without having to add structural reinforcement to the masonry structure.

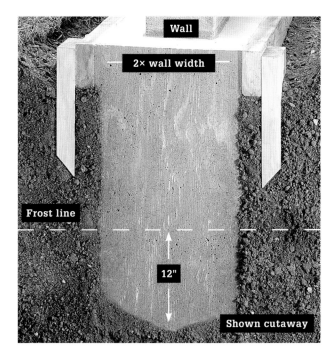

**Frost footings are required** if a structure will be more than 2-ft. tall or if it is tied to another permanent structure. Frost footings should be twice as wide as the structure they support and should extend 8 to 12 inches below the frost line (see pages 42 to 45).

**Low walls** (2 ft. or shorter) can often be built on reinforced concrete slabs—either a newly poured slab or an existing patio surface. Check with your city's building department for specific requirements relating to your type of project.

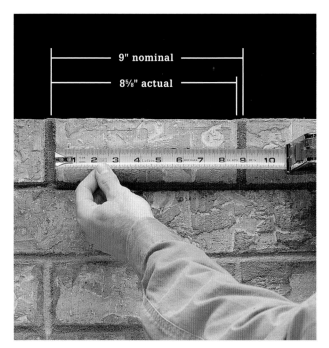

**Do not add mortar joint thickness** to total project dimensions when planning brick and block projects. The actual sizes of bricks and blocks are ⅜" smaller than the nominal size to allow for ⅜"-wide mortar joints. For example, a 9" (nominal) brick has an actual dimension of 8⅝", so a wall that is built with four 9" bricks and ⅜" mortar joints will have a finished length of 36".

**Test project layouts** using ⅜" spacers between masonry units to make sure the planned dimensions work. If possible, create a plan that uses whole bricks or blocks, reducing the amount of cutting required.

# Scoring & Cutting Brick

**Score all four sides of the brick first** with a brick set chisel and maul when cuts fall over the web area and not over the core. Tap the chisel to leave scored cutting marks ⅛ to ¼" deep, then strike a firm final blow to the chisel to split the brick. Properly scored bricks split cleanly with one firm blow.

**When you need to split a lot of bricks** uniformly and quickly, use a circular saw fitted with a masonry blade to score the bricks, then split them individually with a chisel. For quick scoring, clamp them securely at each end with a pipe or bar clamp, making sure the ends are aligned. Remember: wear eye protection when using striking or cutting tools.

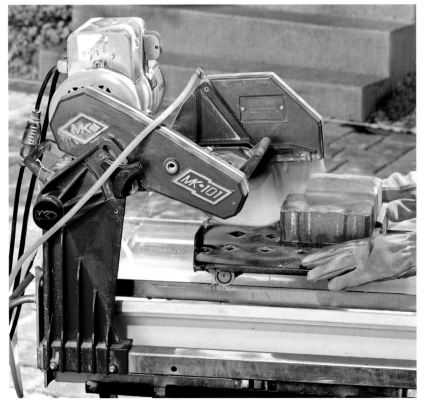

**Masonry saws,** also called wet saws or tub saws, are table-mounted power saws with a circular diamond blade made for cutting a range of clay brick products, tile, and other masonry units. When the saw is running, a sheet of water sprays down over the blade to reduce heat buildup and control dust. To make a cut, set the marked brick on the cutting sled, aligning the mark with the blade. Turn the saw on and slowly push the sled toward the blade until the cut is complete. Masonry saws make clean, accurate cuts and are easy to use. Complete saw setups are commonly available for rent at large home centers and rental outlets.

# How to Use a Brick Splitter

**A brick splitter makes accurate,** consistent cuts in bricks and pavers with no scoring required. It is a good idea to rent one if your project requires many cuts. To use the brick splitter, first mark a cutting line on the brick, then set the brick on the table of the splitter, aligning the cutting line with the cutting blade on the tool.

**Once the brick is in position on the splitter table,** pull down sharply on the handle. The cutting blade on the splitter will cleave the brick along the cutting line. *Tip: For efficiency, mark cutting lines on several bricks at the same time using a drywaller's T-square or a framing square.*

# How to Cut Concrete Block

**Mark cutting lines** on both faces of the block, then score ⅛ to ¼"-deep cuts along the lines using a circular saw equipped with a masonry blade.

**Use a mason's chisel and maul** to split one face of the block along the cutting line. Turn the block over and split the other face.

**Option:** Cut half blocks from combination corner blocks. Corner blocks have preformed cores in the center of the web. Score lightly above the core, then rap with a mason's chisel to break off half blocks.

# Mixing & Placing Mortar

The first critical aspect of working with mortar is getting the mixture right. If it's too thick, it will fall off the trowel in a heap, not in the smooth line that you want. If it's too thin, mortar becomes messy and weak. Always follow the manufacturer's mixing directions, but keep in mind that the amount of water specified is an approximation.

If you've never mixed mortar before, experiment with small amounts until you find a mixture that clings to the trowel just long enough for you to deliver a controlled, even line that holds its shape after settling. Record the best mixture proportions for future batches. Always mix mortar in workable batches, so you'll have time to use it all before it becomes too hard to work. Hot, dry weather shortens the working time. You can add water (called "retempering") to restore workability to hardening mortar, but you must use retempered mortar within two hours.

See page 49 for information about the tools you'll need to complete a successful brick-and-mortar project.

**Standard gray mortar** can be tinted for a wide variety of colors. For best results, add the same amount of colorant to each batch throughout the project. Once you settle on a recipe, record it so you can mix the same proportions each time.

## Working with Mortar

**Test the water absorption rate of bricks** to determine their density. Lower-density bricks can pull too much water from the mortar before it has a chance to set, thus weakening the mortar. Squeeze out 20 drops of water in the same spot on the brick surface. If the surface is completely dry after 60 seconds, dampen the bricks with water before you lay them.

**Buttering bricks** is the technique of applying mortar to the end of the brick before setting it into place. The basic technique is to apply a heavy layer of mortar to one end of the brick, then cut off the excess with a trowel.

# How to Mix & Place Mortar

**Empty mortar mix into a mortar box** and form a depression in the center. Add about ¾ of the recommended amount of water into the depression, then mix it in with a masonry hoe. Do not overwork the mortar. Continue adding small amounts of water and mixing until the mortar reaches the proper consistency. Do not mix too much mortar at one time—mortar is much easier to work with when it is fresh.

**Set a piece of plywood on blocks at a convenient height,** and place a shovelful of mortar onto the surface. Slice off a strip of mortar from the pile, using the edge of your mason's trowel. Slip the trowel point-first under the section of mortar and lift up.

**Snap the trowel gently downward** to dislodge excess mortar clinging to the edges. Position the trowel at the starting point, and "throw" a line of mortar onto the building surface. A good amount is enough to set three bricks. Do not get ahead of yourself. If you throw too much mortar, it will set before you are ready.

**"Furrow" the mortar line** by dragging the point of the trowel through the center of the mortar line in a slight back-and-forth motion. Furrowing helps distribute the mortar evenly.

# Working with Concrete

Concrete work involves three main stages: building forms, mixing and placing the concrete, and finishing the surface. In this book, the projects that use poured concrete are either formed walls or structural footings, both of which require little in the way of skilled finishing work (that's good news, since finishing is the trickiest part of working with concrete). Some basic finishing tips are given on page 41.

Building wall forms is described within each specific project. Pages 42 to 45 show you how to build forms and pour concrete for structural footings. As for mixing the concrete, chances are you'll do that yourself using bags of dry concrete mix. For small projects, you can mix what you need in a wheelbarrow or mortar box, but for larger jobs, it's worthwhile to rent a power mixer (see below). However, if your project calls for at least a cubic yard of concrete, consider ordering a batch of ready-mix from a local supplier. (A cubic yard is 27 cubic feet of volume: for example, a footing that measures one foot wide × four feet deep × nine feet long, or a wall measuring six inches wide × three feet tall × 18 feet long.)

**A good mixture is crucial** to any successful concrete project. Properly mixed concrete is damp enough to form in your hand when you squeeze, and dry enough to hold its shape. If the mixture is too dry, the aggregate will be difficult to work, and will not smooth out easily to produce an even, finished appearance. A wet mixture will slide off the trowel, and may cause cracking and other defects in the finished surface.

## Mixing Concrete On-site

**To mix by hand,** empty entire contents of premixed concrete bags into a mortar box or wheelbarrow. Form a hollow in the mound of mix, and pour in 1 gallon of clean water per 60-lb. bag. Mix with a hoe, adding water in small increments as needed until the right consistency is achieved. Don't overwork the mix. Note the total amount of water used as a reference for the remaining batches.

**To use a power mixer,** fill a bucket with 1 gallon of clean water for each 60-lb. bag of dry concrete mix (most mixers can handle 3 bags at once). Pour in ½ of the water, and then add all of the dry mix. Run the mixer for 1 minute. Add water in small increments as needed to reach the desired consistency, then mix for 3 minutes. Empty the concrete into a wheelbarrow, and rinse the drum immediately.

# Materials for Concrete Projects

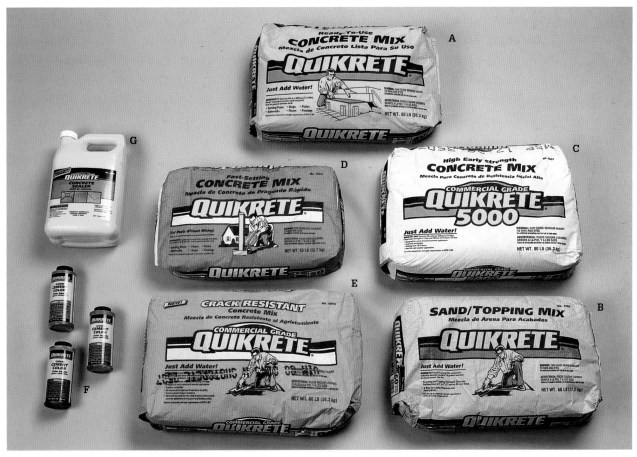

**Concrete mix, usually** sold in 40-, 60- or 80-lb. bags, contains all the components of concrete. You simply add water, mix, and place the concrete. Several varieties are offered at most building centers. The most common are: general-purpose 4,000 psi mix (A), which is the least expensive and is suitable for most do-it-yourself and contractor projects; sand mix (B) contains no large aggregate and is used for shallow pours, such as pouring overlays less than 2" thick (that's why it's sometimes called topping mix); high-early strength mix (C) contains agents and additives that cause it to strengthen quickly, achieving 5,000 psi after 28 days. This mix is particularly appropriate for patios, driveway aprons, and concrete countertops. Other common bagged concrete varieties include fast-setting concrete mix (D) with initial set times under 40 minutes and used for setting posts without mixing; and crack resistant concrete mix (E), a fiber-reinforced concrete mix with improved freeze-thaw durability characteristics.

**Concrete additives include** liquid colorant (F) that is added to the mix to produce vividly colored concrete; and acrylic sealer (G) to promote curing by retaining water in freshly placed concrete.

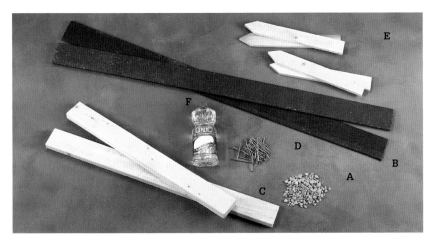

**Materials for subbases** and forms include compactable gravel (A) to improve drainage beneath the poured concrete structure; asphalt-impregnated fiberboard (B) to keep concrete from bonding with adjoining structures and to allow for expansion and contraction of concrete slabs; lumber (C) and 3" screws (D) for building forms; stakes (E) for holding the forms in place; and vegetable oil (F) or a commercial release agent to make it easier to remove the forms.

# Concrete Footings

Footings provide a stable, level base for brick, block, stone, and poured concrete structures. They distribute the weight of the structure evenly, prevent sinking, and keep structures from moving during seasonal freeze-thaw cycles.

The required depth of a footing is usually determined by the frost line, which varies by region. The frost line is the point nearest ground level where the soil does not freeze. In colder climates, it is likely to be 48 inches or deeper. Frost footings (footings designed to keep structures from moving during freezing temperatures) should extend 12 inches below the frost line for the area. Your local building inspector can tell you the frost line depth for your area.

**Footings are required by Building Code for concrete, stone, brick, and block structures** that adjoin other permanent structures or that exceed the height specified by local codes. Frost footings extend 8 to 12" below the frost line. Slab footings, which are typically 8" thick, may be recommended for low, freestanding structures built using mortar or poured concrete. Before starting your project, ask a building inspector about footing recommendations and requirements for your area.

## Tips for Planning ▶

- Describe the proposed structure to your local building inspector to find out whether it requires a footing, and whether the footing needs reinforcement. In some cases, 8"-thick slab footings can be used, as long as the subbase provides plenty of drainage.
- Keep footings separate from adjoining structures by installing an isolation board (see page 44).

# Building Footings

**For poured concrete,** use the earth as a form. Strip sod from around the project area, then strike off the concrete with a screed board resting on the earth at the edges of the top of the trench.

**For brick, block, and stone,** build level, recessed wood forms. Rest the screed board on the frames to create a flat, even surface for stacking masonry units.

**Make footings twice as wide** as the wall or structure they will support. They also should extend at least 12" past the ends of the project area.

**Add tie-rods** if you will be pouring concrete over the footing. After the concrete sets up, press 12" sections of rebar 6" into the concrete. The tie rods will anchor the footing to the structure it supports.

# How to Pour a Footing

**Make a rough outline of the footing** using a rope or hose. Outline the project area with stakes and mason's string. Measure the diagonals or use the 3-4-5 method (page 24) to make sure the string layout is square.

**Strip away sod 6" outside the project area on all sides,** then excavate the trench for the footing to a depth 12" below the frost line.

**Build and install a 2 × 4 form frame for the footing,** aligning it with the mason's strings. Stake the form in place, and adjust to level.

**Variation:** If your project abuts another structure, such as a house foundation, slip a piece of asphalt-impregnated fiber board into the trench to create an isolation joint between the footing and the structure. Use a few dabs of construction adhesive to hold it in place.

**Make two #3 rebar grids to reinforce the footing.** For each grid, cut two pieces of #3 rebar 8" shorter than the length of the footing, and two pieces 4" shorter than the depth of the footing. Bind the pieces together with 16-gauge wire, forming a rectangle. Set the rebar grids upright in the trench, leaving 4" of space between the grids and the walls of the trench. Coat the inside edge of the form with vegetable oil or commercial release agent.

**Mix and pour concrete,** so it reaches the tops of the forms (pages 40 to 41). Screed the surface using a 2 × 4. Float the concrete until it is smooth and level.

**Cure the concrete for one week** before you build on the footing. Remove the forms and backfill around the edges of the footing.

# Fence & Wall Materials

As with most building projects, choosing the right materials for your fence or wall is really a question of priorities. In other words, what do you value most in the finished product: Appearance? Durability? Ease of maintenance? Security? Cost? Ultimately, your decision will involve a combination of priorities (and most likely some compromises). And often the function and style of a fence or wall narrows the choices for you. For example, if you're building a fence for privacy, you can automatically rule out metal fencing. Here's an overview of the most popular fence and wall materials.

## WOOD

Wood is still the most commonly used material for fences and is really the only one that allows for custom designs and details. Durability, cost, and appearance have everything to do with the type and quality of wood you choose. For a painted fence and for structural members (posts and stringers) that aren't highly visible on unpainted fences, the best and cheapest option is pressure-treated (PT) lumber. Unfinished, it doesn't look as good as other wood types, but it's strong and highly rot-resistant, and you can't see it once it's painted. Choose PT lumber

rated for "ground contact" for all posts and any pieces that will be within 6 inches of the ground. Kiln-dried lumber (often labeled KDAT, for "kiln-dried after treatment") is less likely to warp or split than surfaced-dry (S-Dry) lumber.

If you want to stain and seal your fence or leave it unfinished to let it weather naturally, your two standard options are cedar and redwood. Both are naturally rot-resistant, depending on the grade of the lumber. Heartwood (or "all-heart") lumber, which comes from the dense center of the tree, is the most resistant to rot and, in the case of redwood, insects. Sapwood comes from the softer outer portion of the tree and is no more resistant to decay than other softwoods, like pine. Most cedar and redwood lumber you'll find is a mixture of heartwood and softwood, therefore offering varying degrees of limited decay resistance. Discuss your project with knowledgeable staff at a good lumberyard; they can suggest appropriate grades for your project and budget (and the local climate). *Note: When structural strength is important, many fence builders recommend using only PT lumber for all fence posts, due to its superior strength and decay resistance over most cedar and redwood lumber.*

For a small, highly visible and decorative fence, you might consider splurging on a sustainably harvested tropical hardwood, such as ipé, ironwood, meranti, or cambera. Choose these products carefully: the wood should be suitably rot-resistant for your application and local climate, and it should come from a supplier certified for sustainable forestry.

## ORNAMENTAL METAL

Sold in preassembled panels and precut posts made of steel, aluminum, or iron, ornamental metal fencing has a distinctive, formal look reminiscent of traditional wrought iron fences. Most products come prefinished with tough, weather-resistant coatings, making metal fencing one of the lowest-maintenance types you can buy. Steel and aluminum versions are lighter in weight and less expensive than iron fencing and are readily available through home centers and fencing suppliers. Iron fencing made for easy installation is available through specialty manufacturers and distributors.

**Always use galvanized or stainless steel** hardware and fasteners when building fences.

## CHAIN LINK

Chain link is the ultimate utility fence—durable, secure, and virtually maintenance-free. Made of rust-resistant galvanized steel, chain link fencing comes in ready-to-assemble parts and is easily worked into custom lengths and configurations. Installing chain link is a little more involved than with other types of pre-fab fencing, but the technique is pretty straightforward once you get the hang of it.

## WOOD COMPOSITE

Because it's made with wood fibers and plastic, wood composite fencing may be considered an alternative to both wood and vinyl fencing. And it's an environmentally friendly choice to boot. Composite fencing can be made almost entirely from recycled plastic and recycled or reclaimed wood materials (not counting metal brackets and rail stiffeners). Like vinyl, it won't rot and never needs painting. Like wood, it has a solid feel and a non-glare finish. Composite fencing come in ready-to-assemble kits and installs much like many vinyl fence products.

## VINYL

Vinyl fencing is popular for its long life, minimal maintenance requirements (essentially none), and the fact that it comes in many styles based on traditional wood fence designs. As such, vinyl is generally considered an alternative to painted wood fencing. Installation of the various post-and-panel fence systems is relatively simple, provided you follow the manufacturers' instructions carefully.

## BAMBOO

Bamboo occupies its own category because it's not wood—it's grass—and because it has such a unique decorative quality that can turn any fence into a conversation piece. Most bamboo fences are made with preassembled panels (consisting of size-matched canes tied together with wire) set into a wood framework. You can also find materials for building an all-bamboo fence, or you can cover an existing fence with preassembled panels. Bamboo can be tooled and finished with ordinary tools and materials and is an environmentally friendly material.

## FENCE & GATE HARDWARE

All metal hardware and fasteners used for building fences must be corrosion-resistant. This includes hinges, latches, and brackets, as well as screws, nails, and other fasteners. For fastener materials,

**Stone, brick, and block** are timeless, sturdy fence-building materials that stand the test of time.

choose hot-dipped galvanized or stainless steel (not aluminum) when working with pressure-treated wood; with cedar and redwood lumber, galvanized, stainless steel, and aluminum fasteners offer corrosion resistance, but only stainless steel is guaranteed not to discolor the wood.

## STONE

Natural stone is a timeless building material for walls, offering unmatched beauty and durability. Cut stone (called ashlar) is the best choice for most wall applications. Its relatively flat surfaces make it easy to stack for a strong, uniform structure. Other types of stone for building include fieldstone (naturally shaped, irregular stones gathered from fields) and rubble (lower-quality irregular stone pieces used primarily for infill in stone walls).

## BRICK & CONCRETE BLOCK

Clay brick and concrete block are equally well suited to outdoor wall projects. By itself, brick is clearly the more decorative choice, while walls made of block are quicker to build (whether mortared or mortarless) and make a great foundation for decorative finishes like stucco or veneer stone.

## LANDSCAPE BLOCK

Landscape blocks are manufactured concrete units that come in several different forms. All types are uniformly shaped and sized, making them exceptionally easy to work with. In addition to the familiar blocks made specifically for retaining walls, you can now buy building-type landscape blocks designed for do-it-yourself walls, columns, steps, and planters.

# Tools

In addition to basic hand and power tools, your project will likely call for some specialty tools that you can buy at your local home center or a well-stocked hardware store. And with some jobs, heavy-duty power equipment can greatly speed up the work and improve the results. The equipment shown here is commonly available for rent at home centers and rental outlets. In cases where hand tools can be substituted for power equipment, these are mentioned within the given project. Here's a pictorial look at the various specialty tools used in this book, so you'll know exactly what to pick up at the store.

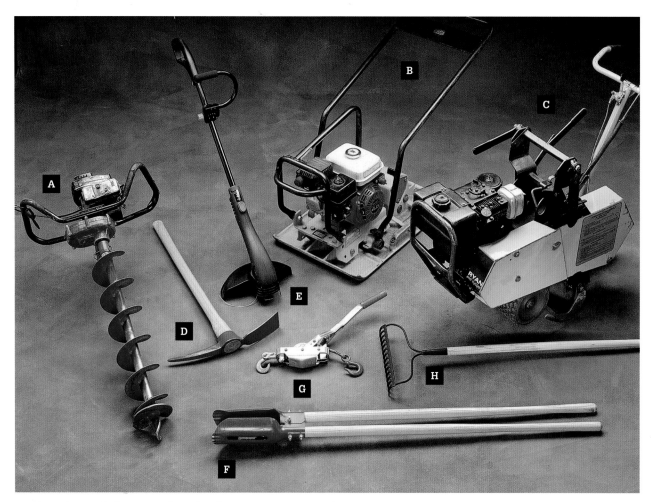

**Landscaping tools for preparing sites for concrete projects include:** power auger (A) for digging holes for posts or poles; power tamper (B) and power sod cutter (C) for driveway and other large-scale site preparation. Smaller landscaping tools include: pick (D) for excavating hard or rocky soil; weed trimmer (E) for removing brush and weeds before digging; posthole digger (F) for when you have just a few holes to dig; come-along (G) for moving large rocks and other heavy objects without excessive lifting; and garden rake (H) for moving small amounts of soil and debris.

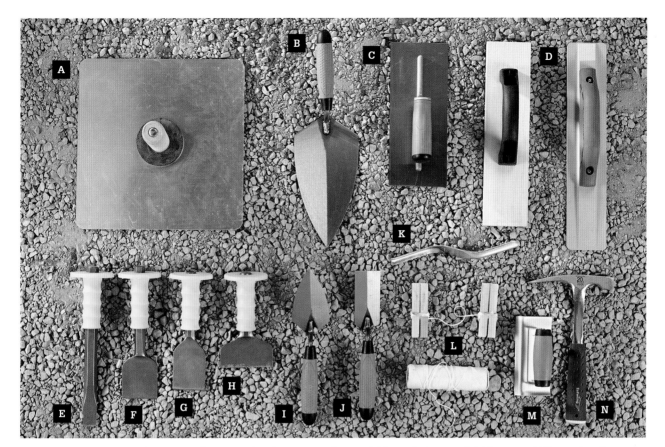

**Concrete, masonry, and stone tools:** Mortar hawk (A), mason's trowel (B), steel concrete finishing trowel (C), wood and magnesium concrete floats (D), pointing chisel (E), pitching chisel (F), stone chisel (G), brickset (H), pointing trowel (I), square-end trowel (J), jointing tool (K), mason's string with line blocks (L), concrete edger (M), mason's hammer (N).

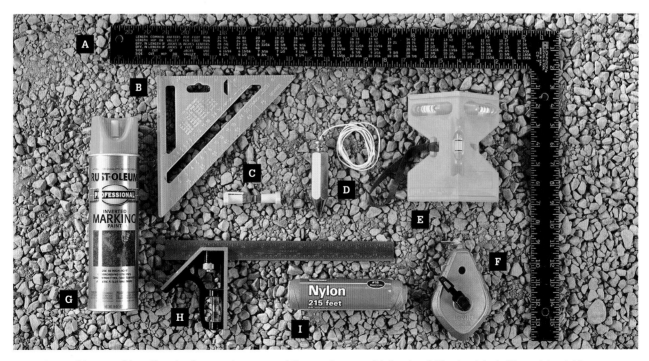

**Layout, marking, and leveling tools:** Framing square (A), speed square (B), line level (C), plumb bob (D), post level (E), chalk line (F), landscape marking paint (G), combination square (H), mason's string (I).

# Fence & Gate Projects

Whether you're a serious weekend warrior or a perfect stranger to construction tools, building your own fence is a worthwhile and doable project. It's outdoor work, it's simple and straightforward, and in most cases it can be done at your own pace. On top of all that, fence building is satisfying work, as you begin to see the fruits of your labor in a very short time.

The projects in this chapter include all of the most popular types of fences and fence materials, as well as several gate options. Many of the fences are custom-built, meaning that you design and construct the fence to your own specifications using stock materials (don't worry, the project steps give you plenty of guidance). You'll also find a range of fences made with prefabricated panels and other modular parts. Wood fencing can be custom or panelized, while vinyl, or wood composite, and all types of metal fencing are strictly prefab.

Before you get started, keep in mind that fence building—like hanging drywall and other jobs involving long, heavy, or unwieldy materials—is often much easier with the help of a friend or two.

## In this chapter:

- Board & Stringer Fence
- Wood Panel Fences
- Picket Fence
- Post & Board Fences
- Split Rail Fence
- Virginia Rail Fence
- Wood Composite Fence
- Vinyl Panel Fence
- Ornamental
  Metal Fence
- Chain Link Fence
  & Gate
- Trellis Fence
- Bamboo Fence
- Invisible Dog Fence
- Brick & Cedar Fence
- Stone & Rail Fence
- Easy Custom Gates
- Arched Gate
- Trellis Gate

# Board & Stringer Fence

The board and stringer (also called vertical board) fence is hands down the most popular type of fence in American suburbs. Why? Because it creates an excellent visual and physical barrier between you and your neighbors. No offense to the neighbors; it's just nice to have a little privacy (especially where house lots are small), not to mention a safe and secure area for the kids to play or the dog to run around in.

The basic structure of a board and stringer fence starts with 4 × 4 or 6 × 6 posts anchored in concrete, typically about eight feet apart. Between each pair of posts, three 2 × 4 stringers (rails) are installed with metal fence brackets or just plain fasteners. The stringers can be positioned on-edge or on the flat; in this project, only the top stringer is set flat. The siding, or infill boards, are fastened one at a time to the stringers and can be fit tightly together (for a solid barrier) or spaced in picket-fence fashion. Another popular style is the staggered board fence, in which the siding boards are installed on both sides of the stringers (which must be on the flat) in alternation. This gives the fence visual depth and provides ventilation while maintaining most of the privacy of a solid barrier.

The piece-built design of this fence makes it a good option for sloping yards. The fence sections can easily be stepped or contoured as desired.

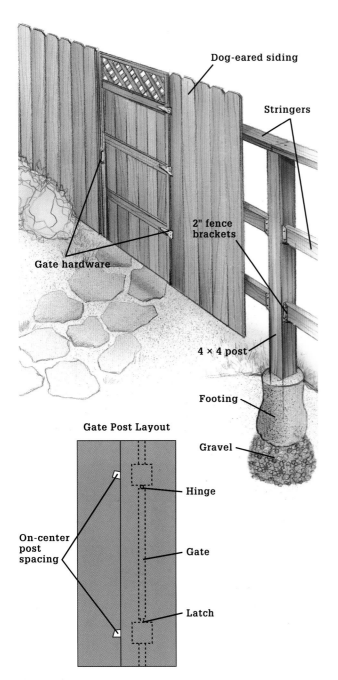

Dog-eared siding

Stringers

2" fence brackets

Gate hardware

4 × 4 post

Footing

Gravel

### Gate Post Layout

Hinge

Gate

On-center post spacing

Latch

## Tools & Materials ▸

| | |
|---|---|
| Tools & materials for setting posts | H.D. galvanized fence bracket nails |
| Tape measure (2) | ⅛" piece of scrap wood |
| Chalk line | Compactable gravel |
| Line level | Hand maul |
| Paintbrush | Hand tamp |
| Circular saw | Reciprocating saw or handsaw |
| Hammer | Plumb bob |
| Drill | Stakes and mason's string |
| Level | Speed square |
| Wood sealer/ preservative | Eye and ear protection |
| Prefabricated gate & hardware | Work gloves |
| Pressure-treated cedar or redwood lumber (4 × 4, 2 × 4, 1 × 6) | Wood blocks |
| | Corrosion-resistant screws |
| Hot-dipped (H.D.) galvanized 2 × 4 fence brackets | Hinge pins |
| | Masking tape |
| H.D. galvanized nails | Pencil |
| H.D. galvanized deck screws | Spray paint |
| | Permanent marker |

# How to Build a Board & Stringer Fence

**1**

**Lay out the fence posts,** spacing them 96" on center, or as desired (see pages 22 to 25). Dig the postholes 6" deeper than the code-required depth. Add 6" of gravel to each hole and tamp it flat. Position each post in its hole, and brace it with 2 × 4 cross bracing so it is perfectly plumb.

**2**

**Check the post alignment with a mason's string.** Pull the string taut and make sure each post touches the line. Make any necessary adjustments to the post positions. Anchor the posts with concrete, checking to make sure they are plumb before the concrete sets (see pages 26 to 29). Let the concrete cure for 48 hours.

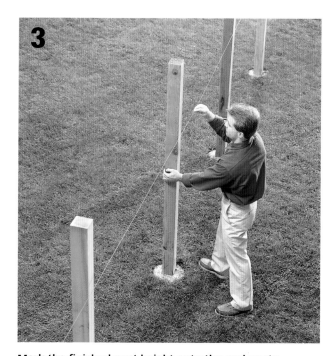

**3**

**Mark the finished post height onto the end posts,** 12" below the planned height of the siding boards. Attach a chalk line to the height marks on the end posts, and snap a cutoff line across the infill posts. (Board and stringer fences are usually constructed so the siding boards extend above the posts.)

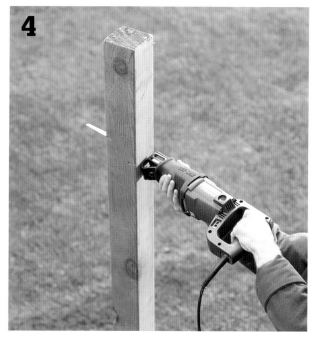

**4**

**Trim off the posts along the marked cutoff lines** using a reciprocating saw or handsaw. Brush sealer-preservative onto the cut ends of the posts.

(continued)

**Cut 2 × 4 top stringers** and coat the ends with sealer-preservative. Center the end joints over the posts, then attach the stringers to the posts with galvanized screws or nails.

**Mark lines on each post** to serve as reference lines for installing additional stringers. Space the marks at 24" intervals, or as desired.

**At each stringer reference mark,** use galvanized nails to attach a 2 × 4 fence bracket to the sides of the posts. Brackets should be flush with the front face of the posts.

**Position a 2 × 4 stringer** between each pair of fence brackets. Hold or tack the stringer against the posts, then mark it for cutting by marking the back side along the edges of the posts. (If the yard is sloped, stringers will be cut at angles.) Cut stringers ¼" shorter than measurement so stringer will slide into brackets easily.

**Slide the stringers into the fence brackets** and attach them with galvanized nails. If stringers are cut at an angle because of the ground slope, bend the bottom flanges on the fence brackets to match this angle before installing the stringers.

**Install the first siding boards,** beginning at an end post. To find the board length, measure from the ground to the top face of the top stringer, and add 10". Cut a number of boards to length. Position the first board so it is 2" above the ground, and use a level to make sure it is perfectly plumb. Fasten the board to the post and stringers with galvanized screws or nails.

**Install the remaining boards,** leaving a gap of at least ⅛" between them (a piece of scrap hardboard or plywood works well as a spacing guide). Check every third or fourth board with the level to make sure it's plumb before fastening. At the corners and ends of the fence, you may need to rip-cut boards to fit.

**Install and finish a prebuilt gate,** as shown in Tip, page 61. Finish the fence with a sealer-preservative or paint, as desired.

# Wood Panel Fences

Prefabricated fence panels take much of the work out of putting up a fence, and (surprisingly) using them is often less expensive than building a board and stringer fence from scratch. They are best suited for relatively flat yards, but may be stepped down on slopes that aren't too steep.

Fence panels come in many styles, ranging from privacy to picket. Most tend to be built lighter than fences you'd make from scratch, with thinner wood for the stringers and siding. When shopping for panels, compare quality and heft of lumber and fasteners as well as cost.

Purchase panels, gate hardware, and gate (if you're not building one) before setting and trimming your posts. Determine also if panels can be trimmed or reproduced from scratch for short sections.

The most exacting task when building a panel fence involves insetting the panels between the posts. This requires that preset posts be precisely spaced and perfectly plumb. In our inset panel sequence (pages 59 to 61), we set one post at a time as the fence was built, so the attached panel position can determine the spacing, not the preset posts.

An alternative installation to setting panels between posts is to attach them to the post faces (pages 62 to 63). Face-mounted panels are more forgiving of preset posts, since the attachment point of stringers doesn't need to be dead center on the posts.

Wood fence panels usually are constructed in either six- or eight-foot lengths. Cedar and pressure-treated pine are the most common wood types used in making fence panels, although you may also find redwood in some areas. Generally, the cedar panels cost one-and-a-half to two times as much for similar styles in PT lumber.

When selecting wood fence panels, inspect every board in each panel carefully (and be sure to check both sides of the panel). These products are fairly susceptible to damage during shipping.

**Building with wood fence panels** is a great time-saver and allows you to create a more elaborate fence than you may be able to build from scratch.

# Tools & Materials ▸

Pressure-treated cedar or redwood 4 × 4 posts
Prefabricated fence panels
Corrosion-resistant fence brackets or panel hangers
Post caps
Corrosion-resistant deck screws (1", 3½")

Prefabricated gate & hardware
Wood blocks
Colored plastic
Tape measure
Plumb bob
Masking tape
Stakes and mason's string
Clamshell digger

or power auger
Gravel
Hand tamp
Level
2 × 4 lumber
Circular saw, hand saw, or reciprocating saw
Concrete
Drill
Line level

Clamps
Scrap lumber
Shovel
Hammer
Speed square
Eye and ear protection
Work gloves
Permanent marker
Hinges (3)

# Panel Board Pattern Variations

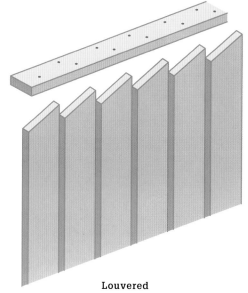

Louvered

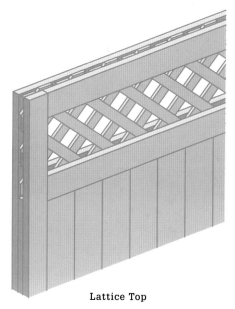

Lattice Top

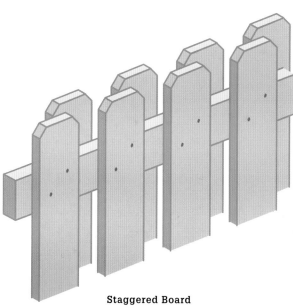

Staggered Board

Stockade

# Installing Fence Panels

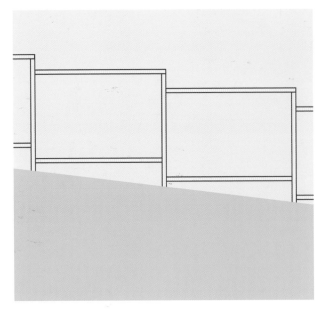

**On a sloped lot,** install the panels in a step pattern, trying to keep a consistent vertical drop between panels. It is difficult to cut most preassembled panels, so try to plan the layout so only full-width panels are used.

**Metal fence panel hangers** make quick work of hanging panels and offer a slight amount of wiggle room if the panel is up to ½" narrower than the space between posts.

**With some panel styles,** the best tactic is to flatten the lower tab after attaching it to the post and then bend it up or down against the panel frame once the panel is in place.

**Setting all of the posts in concrete at one time** and then installing the panels after the concrete sets has advantages as well as disadvantages. On the plus side, this approach lets you pour all of the concrete at the same time and provides good access so you can make absolutely certain the posts are level and plumb. On the downside, if the post spacing is off even a little bit, you'll need to trim the panel (which can be tricky) or attach a shim to the post or the panel frame (also tricky). Most panel manufacturers recommend installing the posts as you go.

# How to Build a Wood Panel Fence

**Lay out the fenceline,** and mark the posthole locations with colored plastic (inset). Space the holes to fit the fence panels, adding the actual post width (3½" for 4 × 4 posts) plus ¼" for brackets to the panel length. *Tip: For stepped fences, measure the spacing along a level line, not along the slope.*

**Dig the first posthole for a corner or end post** using a clamshell digger or power auger. Add 6" of gravel to the hole, and tamp it flat. Set, plumb, and brace the first post with cross bracing.

**Dig the second posthole** using a 2 × 4 spacer to set the distance between posts (cut the spacer to the same length as the stringers on the preassembled fence panels).

**Fill the first posthole with concrete** or with tamped soil and gravel (see pages 26 to 29). Tamp the concrete with a 2 × 4 as you fill the hole. Let the concrete set.

(continued)

**Install the stringer brackets onto the first post** using corrosion-resistant screws or nails. Shorter fences may have two brackets, while taller fences typically have three. *Note: The bottom of the fence siding boards should be at least 2" above the ground when the panel is installed.*

**Set the first panel into the brackets.** Shim underneath the free end of the panel with scrap lumber so that the stringers are level and the panel is properly aligned with the fenceline. Fasten the panel to the brackets with screws or nails.

**Mark the second post for brackets.** Set the post in its hole and hold it against the fence panel. Mark the positions of the panel stringers for installing the brackets. Remove the post and install the stringer brackets, as before.

**Reset the second post,** slipping the ends of the panel stringers into the brackets. Brace the post so it is plumb, making sure the panel remains level and is aligned with the fenceline. Fasten the brackets to the panel with screws or nails.

**9**

**10**

**Anchor the second post in concrete.** After the concrete sets, continue building the fence, following steps 5 to 8. *Option: You can wait to fill the remaining postholes with concrete until all of the panels are in place.*

**Attach the post caps** after trimming the posts to their finished height (use a level mason's line to mark all of the posts at the same height). Install the gate, if applicable; see Tip, below.

## Installing a Prefab Gate ▸

**To install a prefabricated gate,** attach three evenly spaced hinges to the gate frame using corrosion-resistant screws (left). Follow the hardware manufacturer's directions, making sure the hinge pins are straight and parallel with the edge of the gate. Position the gate between the gate posts so the hinge pins rest against one post. Shim the gate to the desired height using wood blocks set on the ground (right). Make sure there is an even gap (reveal) between the gate and the latch post, and then fasten the hinges to the hinge post with screws (inset).

# How to Build a Face-mounted Panel Fence

**Set the posts for your project** (see pages 22 to 29). Since spacing is less critical for face-mounted panels than for inset panels, you can install all of the posts before adding the panels, if desired. Lay out the posts according to the panel size, leaving about ¼" for wiggle room. *Note: Spaces before end, corner, and gate posts must be smaller by half the post width, so that the end of the fence panel covers the entire post face.* Set the posts in concrete.

**Trim the posts to height.** For level or nearly level fences, mark the desired post height on the end posts, allowing for a 2" min. space between the bottom edge of the panels and the ground. Stretch a mason's string between the end/corner posts, and mark the infill posts at the string level. Cut the posts with circular saw, reciprocating saw, or handsaw.

**Position the first panel.** To mark the height for all of the panels, run a mason's string between the end/corner posts to represent the top of the top panel stringers. Use a line level to make sure the line is level. Also make sure the panel will be at least 2" above the ground when installed. Set the first panel onto blocks so the top stringer is touching the mason's string.

**Fasten the first panel.** Holding the panel in position, drill pilot holes and fasten each stringer to the post with 3½" deck screws. Use two screws at each stringer end. At end, corner, and gate posts, the stringers should run all the way across the post faces.

**Install the remaining panels.** Repeat steps 3 and 4 to install the rest of the panels. *Tip: If any posts are off layout, resulting in a stringer joint falling too close to the edge of a post, add a 24"-long brace under the butted stringer ends; the brace should have the same thickness as the stringer stock.* Add post caps and other details, as desired.

**Variation:** Face-mounted privacy fence panels may be fastened to post faces through the panels' vertical frame members. To use this technique, make sure the panel edges are perfectly plumb before fastening, and butt the panels tightly together (or as directed by the manufacturer).

# Picket Fence

The quintessential symbol of American hominess, the classic picket fence remains a perennial favorite for more than its charm and good looks. It's also a deceptively effective boundary, creating a clear line of separation while appearing to be nothing more than a familiar decoration. This unique characteristic of a welcoming barrier makes the picket fence a good choice for enclosing an area in front of the house. It's also a popular option for separating a vegetable or flower garden from the surrounding landscape.

Building a custom picket fence from scratch is a great do-it-yourself project. The small scale and simple structure of the basic fence design make it easy to add your own creative details and personal touches. In this project, you'll see how to cut custom pickets and build a fence using standard lumber (plus an easy upgrade of adding decorative post caps). As an alternative, you can build your fence using prefab fence panels for the picket infill (see pages 59 to 61). You can also buy precut pickets at home centers, lumberyards, and online retailers to save on the work of cutting your own.

Traditionally, a picket fence is about three to four feet tall (if taller than four feet, a picket fence starts to look like a barricade) with 1 × 3 or 1 × 4 pickets. Fence posts can be spaced anywhere up to eight feet apart if you're using standard lightweight pickets. Depending on your preference, the posts can be visible design elements or they can hide behind a continuous line of pickets. Spacing between the pickets is a question of function and taste: go with whatever spacing looks best and fulfills your functional needs.

## Tools & Materials ▸

| | |
|---|---|
| Tools and materials for setting posts | Galvanized or stainless steel finish nails |
| Mason's string | Spacer |
| Line level | Speed square |
| Circular saw | Eye and ear protection |
| Drill | Clamps |
| Power miter saw | Paint brush |
| Sander | Tape measure |
| 2-ft. level | 16d galvanized common nails |
| Lumber (4 × 4, 2 × 4, 1 × 4) | Wood sealant or primer |
| Deck screws (3½, 2") | Work gloves |
| Finishing materials | Pencil |
| Post caps (optional) | Finish materials |
| Hammer | |

**A low picket fence** adds curb appeal and a cozy sense of enclosure to a front yard or entry area without blocking views to or from the house.

## Picket Fence Styles

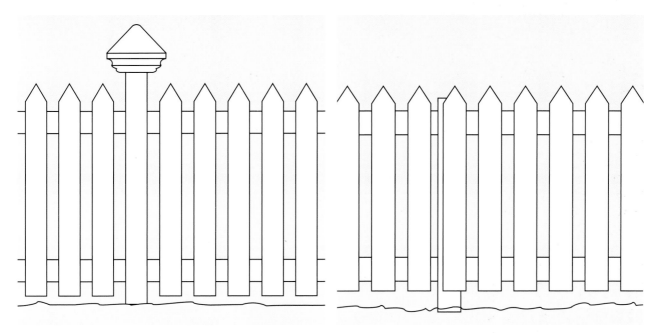

**Highlighting the posts** (left) gives the fence a sectional look, and the rhythm of the pickets is punctuated by the slower cadence of the posts. To create this effect, mount the stringers on edge, so the pickets are flush with—or recessed from—the front faces of the posts. Hiding the posts (right) creates an unbroken line of pickets and a somewhat less structural look overall. This effect calls for stringers installed flush with—or over the front of—the post faces.

## How to Build a Picket Fence

**Install and trim the posts according to your plan** (see pages 22 to 29). In this project, the pickets stand at 36" above grade, and the posts are 38" (without the post caps). Set the posts in concrete, and space them as desired—but no more than 96" on center.

**Mark the stringer positions onto the posts.** Measure down from each post top and make marks at 8 and 28½" (or as desired for your design). These marks represent the top edges of the two stringer boards for each fence section.

(continued)

## Calculating Picket Spacing ▸

Determine the picket quantity and spacing. Cut a few pickets (steps 5 to 7) and experiment with different spacing to find the desired (approximate) gap between pickets. Calculate the precise gap dimension and number of pickets needed for each section using the formula shown in the example here.

Total space between posts: 92.5"

Unit size (picket width + approx. gap size):

    3.5" + 1.75" = 5.25"

Number of pickets (post space ÷ unit size):

    92.5" ÷ 5.25" = 17. 62 (round down for slightly larger gaps; round up for slightly smaller gaps)

Total picket area (# of pickets × picket width):

    17 × 3.5" = 59.5"

Remaining space for gaps (post space - total picket area): 92.5" - 59.5" = 33"

Individual gap size (total gap space ÷ (# of pickets + 1)): 33" ÷ 18 = 1.83"

**Install the stringers.** Measure between each pair of posts, and cut the 2 × 4 stringers to fit. Drill angled pilot holes, and fasten the stringers to the posts with 3½" deck screws or 16d galvanized common nails; drive one fastener in the bottom and top edges of each stringer end.

**Cut the pickets to length** using a power miter saw. To save time, set up a stop block with the distance from the block to blade equal to the picket length. *Tip: If you're painting the fence, you can save money by cutting the pickets from 12-ft.-long boards of pressure-treated lumber. In this project, the pickets are 32" long; each board yields 4 pickets.*

**Shape the picket ends as desired.** For straight-cut designs, use a miter saw with a stop block on the right side of the blade (the first pass cuts through the picket and the block). If the shape is symmetrical, such as this 90° point, cut off one corner, and then flip the board over and make the second cut—no measuring or adjusting is needed.

**Variation:** To cut pickets with decorative custom shapes, create a cardboard or hardboard template with the desired shape. Trace the shape onto each picket and make the cuts. Use a jigsaw for curved cuts. Gang several cut pieces together for final shaping with a sander.

**Prime or seal all surfaces** of the posts, stringers, and pickets; and then add at least one coat of finish (paint, stain, or sealer), as desired. This will help protect even the unexposed surfaces from rot.

**Set up a string line** to guide the picket installation. Clamp a mason's string to two posts at the desired height for the tops of the pickets. *Note: To help prevent rot and to facilitate grass trimming, plan to install the pickets at least 2" above the ground.*

**Install the pickets.** Using a cleat spacer cut to the width of the picket gap, set each picket in place and drill even pairs of pilot holes into each stringer. Fasten the pickets with 2" deck screws. Check the first picket (and every few thereafter) for plumb with a level before piloting.

**Add the post caps.** Wood post caps (with or without metal cladding) offer an easy way to dress up plain posts while protecting the end grain from water. Install caps with galvanized or stainless steel finish nails, or as directed by the manufacturer. Apply the final finish coat or touch-ups to the entire fence.

# Post & Board Fences

Post and board fences include an endless variety of simple designs in which widely spaced square or round posts support several horizontal boards. This type of fence has been around since the early 1700s, when it began to be praised for its efficient use of lumber and land and its refined appearance. The post and board is still a great design today. Even in a contemporary suburban setting, a classic, white three- or four-board fence evokes the stately elegance of a horse farm or the welcoming, down-home feel of a farmhouse fence bordering a country lane.

Another desirable quality of post and board fencing is its ease in conforming to slopes and rolling ground. In fact, it often looks best when the fence rises and dips with ground contours. Of course, you can also build the fence so it's level across the top by trimming the posts along a level line. Traditional agricultural versions of post and board fences typically include three to five boards spaced evenly apart or as needed to contain livestock. If you like the look of widely spaced boards but need a more complete barrier for pets, cover the back side of the fence with galvanized wire fencing, which is relatively unnoticeable behind the bold lines of the fence boards. You can also use the basic post

and board structure to create any number of custom designs. The fence styles shown in the following pages are just a sampling of what you can build using the basic construction technique for post and board fences.

## Tools & Materials ▸

| | |
|---|---|
| Tools and materials for setting posts | 3" stainless steel screws |
| Mason's string | Post levels |
| Line level | Combination square |
| Circular saw | Eye and ear protection |
| Speed square | Lumber (1 × 6, 1 × 4, |
| Clamps | 2 × 6, 1 × 3) |
| Circular saw | Deck screws |
| Drill | (2", 2½", 3½") |
| 4 × 4 posts | 8d galvanized nails |
| Finishing materials | Scrap 2 × 4 |
| Bar clamps | Work gloves |
| Chisel | Pencil |
| Primer paint or stain | |

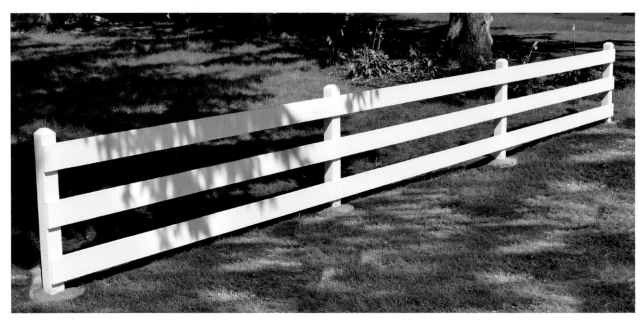

**A low post and board fence,** like traditional picket fencing, is both decorative and functional, creating a modest enclosure without blocking views. The same basic fence made taller and with tighter board spacing becomes an attractive privacy screen or security fence.

# How to Build a Classic Post & Board Fence

**Set the posts in concrete,** following the desired spacing (see pages 22 to 29). Laying out the posts at 96" on center allows for efficient use of lumber. For smaller boards, such as 1 × 4s and smaller, set posts closer together for better rigidity.

**Trim and shape the posts with a circular saw.** For a contoured fence, measure up from the ground and mark the post height according to your plan (post height shown here is 36"). For a level fence, mark the post heights with a level string (see page 29). If desired, cut a 45° chamfer on the post tops using a speed square to ensure straight cuts. Prime and paint (or stain and seal) the posts.

**Mark the board locations** by measuring down from the top of each post and making a mark representing the top edge of each board. The traditional 3-board design employs even spacing between boards. Use a speed square to draw a line across the front faces of the posts at each height mark. Mark the post centers on alternate posts using a combination square or speed square and pencil. For strength, it's best to stagger the boards so that butted end joints occur at every other post (this requires 16-ft. boards for posts set 8-ft. apart). The centerlines represent the location of each butted joint.

(continued)

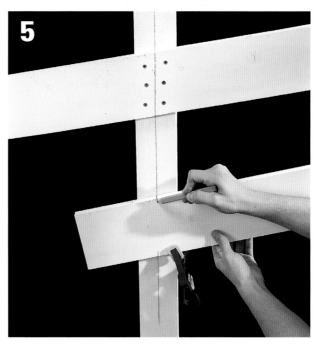

**Install 1 × 6 boards.** Measure and mark each board for length, and then cut it to size. Clamp the board to the posts, following the height and center marks. Drill pilot holes and fasten each board end with three 2½" deck screws or 8d galvanized box nails. Use three fasteners where long boards pass over posts as well.

**Mark for mitered butt joints at changes in elevation.** To mark the miters on contoured fences, draw long centerlines onto the posts. Position an uncut board over the posts at the proper height, and then mark where the top and bottom edges meet the centerline. Connect the marks to create the cutting line, and make the cut. *Note: The mating board must have the same angle for a symmetrical joint.*

**Variation:** This charming fence style with crossed middle boards calls for a simple alteration of the classic three-board fence. To build this version, complete the installation of the posts and top and bottom boards, following the same techniques used for the classic fence. *Tip: If desired, space the posts closer together for steeper cross angles.* Then, mark long centerlines on the posts, and use them to mark the angled end cuts for the middle boards. When installed, the middle boards lap over each other, creating a slight bow in the center that adds interest to the overall look.

# How to Build a Notched Post & Board Fence

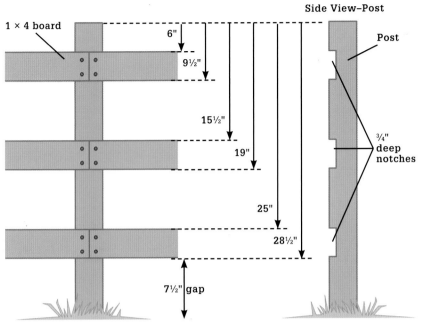

1 × 4 board

6"

9½"

15½"

19"

25"

28½"

7½" gap

Side View–Post

Post

¾" deep notches

**The notched-post fence** presents a slight variation on the standard face-mounted fence design. Here, each run of boards is let into a notch in the posts so the boards install flush with the post faces. This design offers a cleaner look and adds strength overall to the fence. In this example, the boards are 1 × 4s so the posts are set 6 ft. on center; 1 × 6 or 2 × 6 boards would allow for wider spacing (8 ft.). *Note: Because the notches must be precisely aligned between posts, the posts are set and braced before the concrete is added. Alternatively, you can complete the post installation and then mark the notches with a string and cut each one with the posts in place.*

**Cut and mark the posts.** Cut the 4 × 4 posts to length at 66". Clamp the posts together with their ends aligned, and mark the notches at 6, 9½, 15½, 19, 25, and 28½" down from the top ends.

**Create the notches.** Make a series of parallel cuts between the notch marks using a circular saw with the blade depth set at ¾". Clean out the waste and smooth the bases of the notches with a chisel.

**Install the posts and boards.** Set the posts in their holes and brace them in place using a level string to align the notches (see pages 22 to 29). Secure the posts with concrete. Prefinish all fence parts. Install the 1 × 4 boards with 2" deck screws (driven through pilot holes) so their ends meet at the middle of each post.

# How to Build a Capped Post & Board Fence

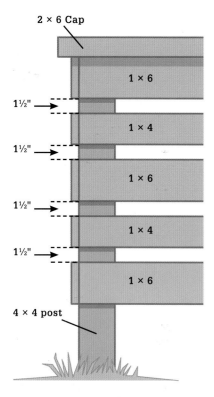

2 × 6 Cap

1 × 6

1½"

1 × 4

1½"

1 × 6

1½"

1 × 4

1½"

1 × 6

4 × 4 post

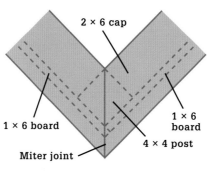

2 × 6 cap

1 × 6 board

1 × 6 board

Miter joint

4 × 4 post

**Top View–Detail**

**A cap rail adds a finished look** to a low post and board fence. This fence design includes a 2 × 6 cap rail and an infill made of alternating 1 × 4 and 1 × 6 boards for a decorative pattern and a somewhat more enclosed feel than you get with a basic 3-board fence. The cap pieces are mitered over the corner posts. Where cap boards are joined together over long runs of fence, they should meet at a scarf joint—made with opposing 30 or 45° bevels cut into the end of each board. All scarf and miter joints should occur over the center of a post.

**Install and mark the posts.** Set the 4 × 4 posts in concrete with 72" on-center spacing (see pages 22 to 29). Trim the post tops so they are level with one another and approximately 36" above grade. Prefinish all fence parts. Use a square and pencil to mark a vertical centerline on each post where the board ends will butt together.

**Install the boards.** For each infill bay, cut two 1 × 4s and three 1 × 6s to length. Working from the top of the posts down, fasten the boards with 2½" deck screws driven through pilot holes. Use a 1½"-thick spacer (such as a 2 × 4 laid flat) to ensure even spacing between boards.

**Add the cap rail.** Cut the cap boards so they will install flush with the inside faces and corners of the posts; this creates a 1¼" overhang beyond the boards on the front side of the fence. Fasten the cap pieces to the posts with 3½" deck screws driven through pilot holes.

# How to Build a Modern Post & Board Privacy Fence

**This beautiful, modern-style post and board fence** is made with pressure-treated 4 × 4 posts and clear cedar 1 × 3, 1 × 4, and 1 × 6 boards. To ensure quality and color consistency, it's a good idea to hand-pick the lumber, and choose S4S (surfaced on four sides) for a smooth, sleek look. Alternative materials include clear redwood, ipé, and other rot-resistant species. A high-quality, UV-resistant finish is critical to preserve the wood's natural coloring for as long as possible.

**Install the posts,** spacing them 60" on-center (see pages 22 to 29) or as desired. Mark the tops of the posts with a level line, and trim them at 72" above grade. *Note: This fence design is best suited to level ground.* Cut the fence boards to length. If desired, you can rip down wider stock for custom board widths (but you'll have to sand off any saw marks for a finished look).

**Fasten the boards to the post faces** using 2½" deck screws or 8d galvanized box nails driven through pilot holes. Work from the top down, and use ⅞"-thick wood spacers to ensure accurate spacing.

**Add the battens to cover the board ends** and hide the posts. Use 1 × 4 boards for the infill posts and 1 × 6s for the corner posts. Rip ¾" from the edge of one corner batten so the assembly is the same width on both sides. Fasten the battens to the posts with 3" stainless steel screws (other screw materials can discolor the wood).

# Split Rail Fence

The split rail, or post and rail, fence is essentially a rustic version of the post and board fence style (pages 68 to 73) and is similarly a good choice for a decorative accent, for delineating areas, or for marking boundaries without creating a solid visual barrier. Typically made from split cedar logs, the fence materials have naturally random shaping and dimensions, with imperfect details and character marks that give the wood an appealing hand-hewn look. Natural weathering of the untreated wood only enhances the fence's rustic beauty.

The construction of a split rail fence couldn't be simpler. The posts have holes or notches (called mortises) cut into one or two facets. The fence rails have trimmed ends (called tenons) that fit into the mortises. No fasteners are needed. Posts come in three types to accommodate any basic configuration: common posts have through mortises, end posts have half-depth mortises on one facet, and corner posts have half-depth mortises on two adjacent facets. The two standard fence styles are two-rail, which stand about three feet tall, and three-rail, which stand about four feet tall. Rails are commonly available in eight- and ten-feet lengths.

In keeping with the rustic simplicity of the fence design, split rail fences are typically installed by setting the posts with tamped soil and gravel instead of concrete footings (frost heave is generally not a concern with this fence, since the joints allow for plenty of movement). This comes with a few advantages: the postholes are relatively small, you save the expense of concrete, and it's much easier to replace a post if necessary. Plan to bury about a third of the total post length (or 24 inches minimum). This means a three-foot-tall fence should have 60-inch-long posts. If you can't find long posts at your local home center, try a lumberyard or fencing supplier.

## Tools & Materials ▶

| | |
|---|---|
| Mason's string | Precut split rail fence |
| Shovel | posts and rails |
| Clamshell digger | Compactable gravel |
| or power auger | (bank gravel |
| Digging bar | or pea gravel) |
| (with tamping | Plastic tags |
| head) or 2 × 4 | Lumber and screws for |
| Level | cross bracing |
| Reciprocating saw | Wheelbarrow |
| or handsaw | Line level |
| Tape measure | Shovel |
| Stakes | Eye and ear protection |
| Soil | Work gloves |
| Nails | |

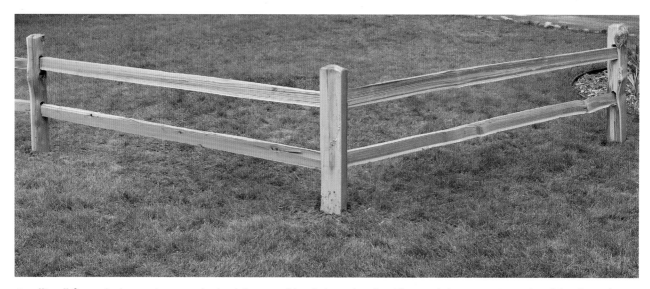

**A split rail fence** looks great as a garden backdrop or a friendly boundary line. The rough-hewn texture and traditional wood joints are reminiscent of homesteaders' fences built from lumber cut and dressed right on the property.

# How to Build a Split Rail Fence

**Determine the post spacing** by dry-assembling a fence section and measuring the distance between the post centers. Be sure the posts are square to the rails before measuring.

**Set up a string line** using mason's string and stakes to establish the fence's path, including any corners and return sections. Mark each post location along the path using a nail and plastic tag.

**Dig the postholes** so they are twice as wide as the posts and at a depth equal to ⅓ the total post length plus 6". Because split posts vary in size, you might want to lay out the posts beforehand and dig each hole according to the post size.

**Add 6" of drainage gravel to each posthole.** Tamp the gravel thoroughly with a digging bar or a 2 × 4 so the layer is flat and level.

(continued)

**5**

**Set and measure the first post.** Drop the post in its hole, and then hold it plumb while you measure from the ground to the desired height. If necessary, add or remove gravel and re-tamp to adjust the post height.

**6**

**Brace the post with cross bracing** so it is plumb. Add 2" of gravel around the bottom of the post. Tamp the gravel with a digging bar or 2 × 4, being careful not to disturb the post.

**7**

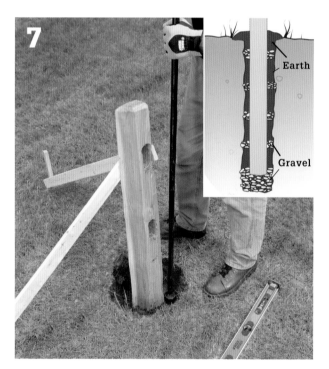

Earth

Gravel

**Fill and tamp around the post,** one layer at a time. Alternate between 4" of soil and 2" of gravel (inset), tamping each layer all the way around the post before adding the next layer. Check the post for plumb as you work. Overfill the top of the hole with soil and tamp it into a hard mound to help shed water.

**8**

**Assemble the first section of fence** by setting the next post in its hole and checking its height. Fit the rails into the post mortises, and then brace the second post in place. *Note: Set all the posts at the same height above grade for a contoured fence. For a level fence, see Variation, right.*

**Variation:** For a fence that remains level across the top, set up a level mason's line strung between two installed fence posts or between temporary supports. Set all of the posts so their tops are just touching the line.

**Secure the second post** by filling and tamping with alternate layers of gravel and soil, as with the first post. Repeat steps 5 through 9 to complete the fence. *Tip: Set up a mason's string to help keep the posts in a straight line as you set them.*

## Custom Details ▸

**Custom-cut your rails** to build shorter fence sections. Cut the rails to length using a reciprocating saw and long wood blade or a handsaw (be sure to factor in the tenon when determining the overall length). To cut the tenon, make a cardboard template that matches the post mortises. Use the template to mark the tenon shape onto the rail end, and then cut the tenon to fit.

**Gates for split rail fences** are available from fencing suppliers in standard and custom-order sizes. Standard sizes include 4 ft. for a walk-through entrance gate and 8 or 10 ft. for a drive-through gate. For large gates, set the side posts in concrete footings extending below the frost line.

# Virginia Rail Fence

The Virginia Rail fence—also called a worm, snake, and zigzag fence—was actually considered the national fence by the U.S. Department of Agriculture prior to the advent of wire fences in the late 1800s. All states with farmland cleared from forests had them in abundance. The simplest fences were built with an extreme zig-zag, and didn't require posts. To save on lumber and land, farmers began straightening the fences and burying pairs of posts at the rail junctures.

Feel free to accommodate the overlapping rail fence in this project to suit your tastes and needs. Increase the zig-zag to climb rolling ground, decrease it to stretch the fence out. Lapped rail fences should be made from rot resistant wood, like cedar, locust, or cyprus.

For the most authentic-looking fence, try to find split, rather than sawn, logs. For longevity, raise the bottom rail off the ground with stones. Posts may eventually rot below ground, but the inherently stable zig-zag form should keep the fence standing until you can replace them.

## Tools & Materials ▶

Mason's string
Shovel
Clamshell digger
Digging bar
Pliers or wire cutters
Large screwdriver
Sledgehammer
Reciprocating saw
    or handsaw
Stakes
Split cedar fence
    posts and rails
Marking paint
Stones
Clothesline or rope

9-gauge
    galvanized wire
Wood blocks
Tape measure
Level
Scrap wood
Eye and ear
    protection
Work gloves

**The Virginia Rail fence** exhibits a very familiar style to anyone who has spent much time in countryside that was cleared and farmed in the 18th and 19th centuries. Since nails were scarce, these zig-zagging post and rail fences were popular because they are held together with only wire or rope.

# How to Build a Virginia Rail Fence

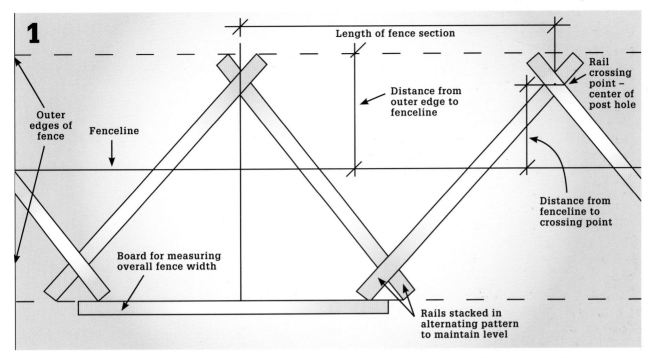

**1**

Length of fence section

Rail crossing point – center of post hole

Distance from outer edge to fenceline

Outer edges of fence

Fenceline

Distance from fenceline to crossing point

Board for measuring overall fence width

Rails stacked in alternating pattern to maintain level

**Plan the layout of your fence** by setting down three or four sections formed with single rails without posts. Set the rails over a mason's string, rope, or garden hose that represents the fenceline running down the middle of the fence. Experiment with different angles of zigzag: more acute angles create a more stable fence over rolling contours, but this requires more lumber and takes up more space. Also determine how much overlap you want at the rail ends. When you are satisfied with your layout, use a board spanning across the open side of a fence section to measure the overall width, or path, of the fence.

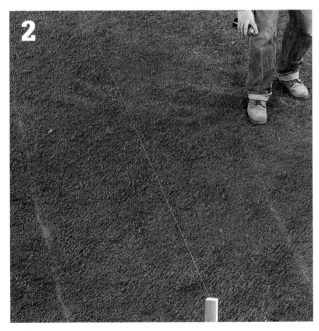

**Mark the fenceline with stakes and mason's string.** This will become the center of the fence's path. Then, using the measurements taken from your layout, mark the ground on either side of the fenceline to represent the outer edges of the fence path. Use marking paint or mason's strings to mark the edges.

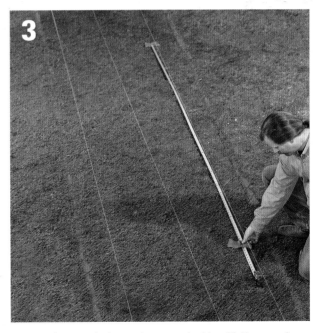

**Locate the posthole centers** to coincide with the crossing points of the rails. To keep the posts in line, plot the postholes along additional mason's strings representing the distance from the fenceline to the rail crossing points. Make sure that the posts alternate from side to side with even spacing throughout the fence run.

(continued)

**Dig the first pair of postholes** using a clamshell digger. Make the holes about three times the width of one post and 18 to 24" deep. Because of the fence's inherent flexibility, the posts don't need to extend below the frost line.

**Place two posts in each hole,** leaving enough room for a rail to pass in between them. Hold the posts plumb, and backfill the holes with soil, compacting it moderately to allow for some movement of the posts, if necessary.

**Thread a rail through the post pairs,** propping it up near the ends with rocks or landscape blocks. Cinch the top ends of the posts together with clothesline or rope to keep them parallel. The rail should extend past the posts an equal distance at both ends.

**Continue building the fence in the same fashion.**
Remember to alternate the rail placement to keep the rails roughly level. You can use chunks of scrap wood from the rail or post material as spacers to help level uneven rails, if desired.

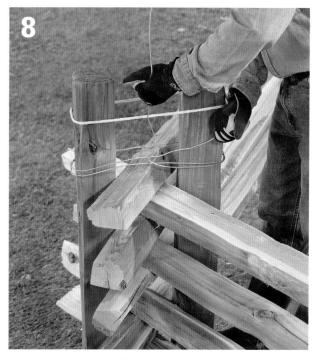

**Bind the post pairs with wire** once the fence sections are complete. Wrap 9-gauge galvanized wire a couple of times around the post, directly above the top-most rail. Twist the wire ends together a couple of times, leaving longish tails after the twist.

**Tighten the wire with a screwdriver.** Twist the wire tails around the shaft of a large screwdriver a few times, and then rotate the screwdriver in a circle (as if you're applying a tourniquet) until the wire is tight and begins to bite into the posts. Trim the wire tails and twist them under or drive them into the wood for safety.

**Drive the posts in further,** if necessary, to stiffen up the post and rail junctures. Protect the tops of the post with a wood block. You can also tamp around the posts with a digging bar to stabilize them. If necessary, trim the post tops of each pair so they are even.

# Wood Composite Fence

Wood composite fencing requires little maintenance and can last a lifetime. For many homeowners, this low-maintenance longevity justifies the high initial cost of the fencing. Manufacturers of composite products claim that they are less expensive than wood in the long run, when you factor in the repair, refinishing, and eventual replacement of wood fences over the years. Quality composite fencing is guaranteed for up to 25 years not to split, crack, splinter, or rot. Perhaps best of all, it never needs to be painted or sealed for protection from the elements.

Composite fences are made from a blend of wood fibers and plastic resins and can contain a high percentage of recycled materials (the country's largest manufacturer of wood composite products uses seven out of every 10 grocery bags recycled nationally). Most of the wood used comes from reclaimed sawdust from woodworking industries and discarded shipping pallets. The reuse of waste materials, combined with the fact that the fencing never needs to be finished and may never need to be replaced, makes wood composite one of the most environmentally friendly fence materials available.

Like vinyl fencing, composite systems are assembled from precisely manufactured components and panels. This makes it difficult to modify the length of fence sections, should your post spacing be off. For this reason, you might prefer to set the posts as you go (instead of all at once), using a fence stringer to determine the exact post placement. If your site is sloped, check with the fencing manufacturer for recommendations on stepping or contouring the fence to follow the slope.

## Tools & Materials ▸

| | |
|---|---|
| Supplies for laying out and setting posts | Composite fence materials and hardware |
| Drill | Galvanized finish nails or adhesive |
| Circular saw and carbide-tipped wood blade | Eye and ear protection |
| Hacksaw | Hammer |
| Level | Work gloves |

**Composite fencing** is manufactured with a blend of wood fibers and plastic resins. It is denser than vinyl fencing and available in a wide range of colors and textures; some even replicate the look of real wood. The privacy fence above is from the Seclusions line by Trex.

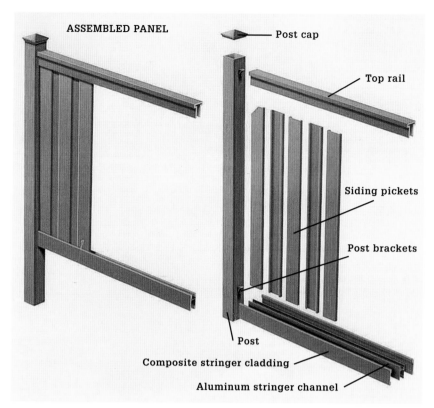

ASSEMBLED PANEL

Post cap

Top rail

Siding pickets

Post brackets

Post

Composite stringer cladding

Aluminum stringer channel

**The composite fence system** shown in this project is supported by hollow composite posts set in concrete. Bottom stringers include aluminum channels (for strength) clad with composite sleeves. The siding infill is made up of interlocking pieces (or pickets) that fit into the bottom stringer channels and are covered with a top rail. The bottom stringer and top rail are anchored to the posts with brackets.

# How to Construct a Wood Composite Fence

**Dig the postholes according to your fence layout** (see pages 22 to 25 for general layout steps). Be sure to follow the manufacturer's directions for post spacing. Dig the holes 12" in diameter and 30" deep (or as directed). Add 6" of gravel to each hole and tamp it flat.

**Set the posts in concrete** using a layout string to ensure precise alignment of the post faces (see pages 26 to 29). Brace each post with cross bracing so it is perfectly plumb. Fill around the post with concrete, up to 2" below ground level. Tamp the concrete with a 2 × 4 to eliminate air pockets. Let the concrete cure for 24 to 48 hours.

(continued)

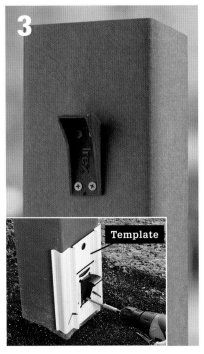

**Install the post brackets** with the provided screws, using the manufacturer's template (inset) to center the brackets on the post faces. Install the bottom bracket onto each post first, measure up from the bracket and mark the position of the top bracket, and then install the top bracket.

**Assemble each bottom stringer** by sliding the composite cladding pieces over the sides of the aluminum stringer channel. For short fence sections, see step 5.

**Cut a stringer as needed** for short sections of fence. Cut the aluminum channel with a hacksaw. Trim the composite cladding pieces to match the channel with a circular saw and carbide-tipped wood blade.

**Set the stringer** onto the bottom post brackets. Check the stringer with a level. If necessary, remove the stringer and adjust the bracket heights (you may have to adjust top brackets as well to maintain the proper spacing).

**Fasten the stringer** ends to the post brackets using the provided screws.

**Trim the upper outside corner** of the first picket so it will clear the top post bracket using a circular saw and carbide-tipped wood blade.

**Install the first picket** by slipping its bottom end into the stringer channel. Align the picket to the top post bracket, and fasten the picket to the post with three evenly spaced screws.

**Assemble the fence panel** by fitting the pickets together along their interlocking side edges and sliding their bottom ends into the stringer channel.

**Fit the last picket into place** after trimming its top corner to clear the post bracket, as you did with the first panel. Fasten the picket to the post with three screws, as in step 9.

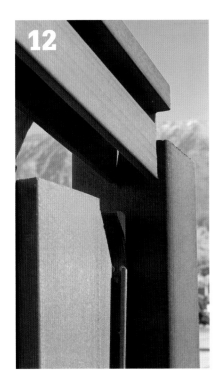

**Set the top rail** over the ends of the pickets until the rail meets the top post brackets.

**Secure the top rail** to each top post bracket, using the provided screws, driving the screws through the top of the rail and into the bracket.

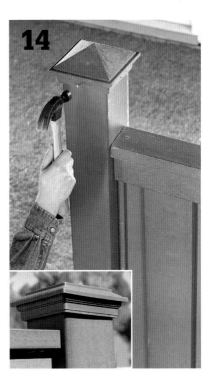

**Add the post caps,** securing them to the posts with galvanized finish nails or an approved adhesive. *Tip: Some fence manufacturers offer more than one cap style (inset).*

# Vinyl Panel Fence

The best features of vinyl fencing are its resilience and durability. Vinyl fencing is made with a form of tough, weather-resistant, UV-protected PVC (polyvinyl chloride), a plastic compound that's found in numerous household products, from plumbing pipe to shower curtains. A vinyl fence never needs to be painted and should be guaranteed for decades not to rot, warp, or discolor. So if you like the styling of traditional wood fences, but minimal maintenance is a primary consideration, vinyl might just be your best option. Another good option is wood composite fencing (see pages 82 to 85), which comes in fewer styles than vinyl but is environmentally friendly and can replicate the look of wood fencing.

Installing most vinyl fencing is similar to building a wood panel fence. With both materials, it's safest to set the posts as you go, using the infill panels to help you position the posts. Accurate post placement is critical with vinyl, because many types of panels cannot be trimmed if the posts are too close together. Squeezing the panel in can lead to buckling when the vinyl expands on hot days, while setting the posts too far apart results in unsightly gaps.

Given the limited workability of most vinyl panels, this fencing tends to work best on level or gently sloping ground. Keep in mind that installation of vinyl fences varies widely by manufacturer and fence style.

## Tools & Materials ›

| | |
|---|---|
| Mason's string | Vinyl fence materials |
| Shovel | (with hardware, |
| Clamshell digger | fasteners, and |
| or power auger | decorative accessories) |
| Circular saw | Pea gravel |
| Drill | Concrete |
| Tape measure | Pressure-treated 4 × 4 |
| Hand maul | (for gate, if applicable) |
| Line level | PVC cement or screws |
| Post level | (optional) |
| Clamps or duct tape | Work gloves |
| Concrete tools | Post caps |
| Stakes | Eye and ear protection |
| 2 × 4 lumber | |

**Vinyl fencing** is now available in a wide range of traditional designs, including picket, post and board, open rail, and solid panel. Color options are generally limited to various shades of white, tan, and gray.

# How to Install a Vinyl Panel Fence

**Lay out the first run of fence** with stakes and mason's string. Position the string so it represents the outside or inside faces of the posts (you'll use layout strings to align the posts throughout the installation). Mark the center of the first post hole by measuring in from the string half the post width.

**Dig the first posthole,** following the manufacturer's requirements for diameter and depth (improper hole dimensions can void the warranty). Add 4 to 6" (or as directed) of pea gravel to the bottom of the hole and tamp it down so it is flat and level using a 2 × 4 or 4 × 4.

**Attach the fence panel brackets to the first post** using the provided screws. Dry-fit a fence panel into the brackets, then measure from the top of the post to the bottom edge of the panel. Add 2" (or as directed) to represent the distance between the fence and the ground; the total dimension is the posts' height above the ground.

**Set up a post-top string** to guide the post installation. Using the post height dimension, tie a mason's string between temporary 2 × 4 supports so the string is centered over the post locations. Use a line level to make sure the string is level. Measure from the string to the ground in several places to make sure the height is suitable along the entire fence run.

(continued)

**Set the first post.** Drop the post in its hole and align it with the fenceline string and height string. Install cross bracing to hold the post perfectly plumb. *Tip: Secure bracing boards to the post with spring-type clamps or duct tape.* Fill the posthole with concrete and let it set completely.

**Determine the second post's location** by fitting a fence panel into the brackets on the first post. Mark the ground at the free edge of the panel. Measure out from the mark half the post width to find the center of the post hole (accounting for any additional room needed for the panel brackets.)

**Complete the fence section.** Dig the hole for the second post, add gravel, and tamp as before. Attach the panel brackets to the second post, set the post in place and check its height against the string line. Assemble the fence section with the provided screws (inset). Confirm that the fence panel is level. Brace the second post in place (as shown) and anchor it with concrete. Repeat the same layout and construction steps to build the remaining fence sections.

## Cutting Panels ▶

**Cut panels for short runs** on solid-panel fencing (if straight along the top) per manufacturer's recommendations.

## 8

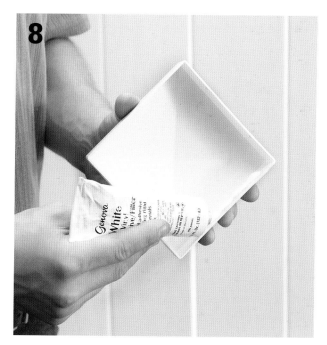

**Add the post caps.** Depending on the product, caps may be installed with PVC cement or screws, or they may be fitted without fasteners. Add any additional decorative accessories, such as screw caps, to complete the installation.

# Installing a Vinyl Fence Gate

**Hang the gate** using the provided hardware. Fasten the hinges to the gate panel with screws. Position the gate in line with the infill fence panels, and screw the hinges to the hinge post. Install the latch hardware onto the gate and latch post. Close the gate, position the gate stops against the gate rails, and fasten the stops to the latch post with screws.

## Post Infills ▶

Reinforce the hinge post with a pressure-treated 4 × 4 inserted inside the post. Set the post in concrete following the same steps used for fence sections. Check carefully to make sure the post is plumb, as this will ensure the gate swings properly. Install the latch post according to the manufacturer's specified dimension for the gate opening.

# Ornamental Metal Fence

Ornamental metal fencing is so called to distinguish it from the other common metal fence material, chain link, which makes a useful fence, but is far from ornamental. Ornamental metal fences arguably offer the best combination of strength, durability, and visibility of any standard fence type. In general, most ornamental metal fences are modern iterations of traditional iron, or "wrought iron," fencing and offer a similarly elegant, formal look (if perhaps not the same heft and handcrafted character).

Today, most ornamental metal fencing is made with galvanized steel or aluminum. Both are finished with durable powder coatings for weather resistance, and most fence systems are based on modular components designed for easy DIY installation. Comparing the two materials, appearances are virtually identical, while aluminum is lighter in weight. It also tends to carry a longer warranty than steel products, probably because aluminum is a naturally rust-proof material. The other type of ornamental fence is iron, which is available in a variety of forms, including bolt-together modular systems (see page 93).

Thanks to its exceptional security and visibility, ornamental metal fencing is a very popular choice for upscale yards. That's why most manufacturers offer gates (with welded construction for strength) and code-compliant locking hardware as standard options. Some fence lines include special infill panels and gates with closer picket spacing than standard panels. If you're installing your fence as a pool surround, check the local codes for requirements.

## Tools & Materials ▸

| | |
|---|---|
| Mason's string | Lumber and screws |
| Tape measure | for cross bracing |
| Shovel | Wheelbarrow |
| Clamshell digger | Masking tape |
| or power auger | Marking paint |
| Clamps or duct tape | Level |
| Drill | Hacksaw |
| Concrete tools | Drainage gravel |
| Post level | Eye and ear |
| Stakes | protection |
| 2 × 4 or 4 × 4 lumber | Work gloves |
| Modular fence materials | 1 × 3 or 1 × 4 lumber |
| Concrete | Permanent marker |

**Ornamental steel, aluminum, and iron fences** come in prefabricated panels up to 6 ft. in height and 8 ft. in length, with matching posts and optional decorative details. The most common color option is black (the better to mimic the look of wrought iron), but some products come in white, bronze, and other colors.

# How to Install an Ornamental Metal Fence

**Lay out the fenceline** with stakes and mason's string. Start at the corners, driving stakes a few feet beyond the actual corner so that the strings intersect at 90° (as applicable). Mark the approximate post locations onto the strings using tape or a marker.

**Mark the first post location** with ground-marking spray paint. Assemble the panel onto the first post and align it in the corner with the mason's strings.

**Dig the first posthole,** following the manufacturer's specifications for depth and diameter. Shovel drainage gravel into the hole, and tamp it with a 2 x 4 or 4 x 4. Set the post in the hole and measure its height above the ground. If necessary, add or remove gravel until the post top is at the recommended height.

**Plumb and anchor the first post.** Position the post perfectly plumb using a post level. Brace the post with cross bracing. Use clamps to secure the bracing to the post. Fill the hole with concrete and let it set.

**Drill pilot holes** for the brackets into the second post and first panel. Align the fence assembly with the first post and mark for the second post hole. Prepare the post hole as you did in step 3.

**6**

**Fill the second posthole with concrete and let it set.**
Here, we have a temporary brace to hold the post plumb and at the desired height. The first panel, complete with posts on either sides, is now set. Remaining posts along this fenceline can be set by positioning posts with spacers to save time (see tip on this page).

**7**

**Align the second panel** on the other side of the corner post. Follow instructions in steps 2 through 5 to set the post and install the panel. Repeat the same process to install the remaining fence sections. You can save time by positioning the posts with spacers (see tip on this page).

**Variation:** For brick pillar corners, columns, or the side of a house, install manufacturer-provided wall brackets. If wall brackets do not come with the standard installation package of your metal fence, contact the manufacturer.

## Spacing Posts ▶

Spacers help you locate the posts without having to measure or install each panel for every post. The panels are then added after the post concrete has set. Create each spacer with two 1 × 3 or 1 × 4 boards. Cut board(s) to fit flush from outside edge to outside edge of the first and second post (once they are set in concrete). Clamp the board in between an anchored post and the next post to be installed. It is best to position spacer boards near the top and bottom of the posts. With the boards in place, the linear spacing should be accurate, but always check the new post with a level to make sure it is plumb before setting it in concrete. Use a level mason's string to keep the post brackets at the same elevation.

# How to Cut Metal

**1**

**Measure and mark panels for cuts.** Hold the panel up to the final post in the run and mark the cutting line. Often, designs will not accommodate full panels around the entire fence perimeter.

**2**

**Cut panels to the appropriate length** using a hacksaw, as needed.

## Old (and Old-fashioned) Iron Fencing ▶

Traditional iron fencing—commonly called "wrought iron"—has been adorning and securing homes and other buildings for many centuries and is still the gold standard of ornamental metal fencing. The oldest forms of wrought iron fences were made with individually hand-forged pieces, while cast-iron fences were assembled from interchangeable pieces of molded iron. Wrought iron, the material, is a pure form of iron that contains very little carbon. Most modern iron fences are made of a form of steel, not wrought iron.

While new iron fencing can still be made by the hand of a blacksmith, it's also commonly available in preassembled panels and modular posts, much like the steel and aluminum fencing sold at home centers. Some iron fencing must be welded together on-site (by professional installers), while some is assembled with bolts, making it suitable for DIY installation. Many styles of prefab iron fencing can be surprisingly affordable.

If you have your heart set on the timeless look and feel of iron, search online for local fabricators and dealers of real iron fencing. You can also hunt through local architectural salvage shops, where you can find antique iron fence panels, posts, finials, and other adornments. Their condition may not be perfect, but the patina of weathering and marks of use only add to the character of old iron fencing.

**Whether it was made yesterday or in the 1800s,** iron fencing offers enduring beauty and unmatched durability, making it worth the splurge on a small fence or a front entry gate.

# Chain Link Fence & Gate

If you're looking for a strong, durable, and economical way to keep pets and children in—or out—of your yard, a chain link fence may be the perfect solution. Chain link fences require minimal maintenance and provide excellent security. Erecting a chain link fence is relatively easy, especially on level property. Leave contoured fence lines to the pros. For a chain link fence with real architectural beauty, consider a California-style chain link with wood posts and rails (see pages 99 to 100).

A 48-inch-tall chain link fence—the most common choice for residential use—is what we've demonstrated here. The posts, fittings, and chain link mesh, which are made from galvanized metal, can be purchased at home centers and fencing retailers. The end, corner, and gate posts, called terminal posts, bear the stress of the entire fence line. They're larger in diameter than line posts and require larger concrete footings. A footing three times the post diameter is sufficient for terminal posts. A properly installed stringer takes considerable stress off the end posts by holding the post tops apart.

When the framework is in place, the mesh must be tightened against it. This is done a section at a time with a winch tool called a come-along. As you tighten the come-along, the tension is distributed evenly across the entire length of the mesh, stretching it taut against the framework. One note of caution: it's surprisingly easy to topple the posts if you over-tighten the come-along. To avoid this problem, tighten just until the links of the mesh are difficult to squeeze together by hand.

Instructions for installing a chain link gate are given on page 98. If you're building a new fence, it's a good idea to test-fit the gate to make sure the gate posts are set properly before you complete the fence assembly.

## Tools & Materials ▸

| | |
|---|---|
| Supplies for setting posts | 3" deck screws or 16d galvanized common nails |
| Mason's string | Post finials or caps |
| Ratchet wrench | Tension wire |
| Pliers | Large galvanized fence staples |
| Hacksaw or pipe cutter | Hog rings |
| Chain link fence materials and hardware | Lumber for cross bracing |
| Duct tape | Level |
| Tie wire | Permanent marker |
| Circular saw, reciprocating saw, or handsaw | Speed square |
| Drill | Clamps |
| Come-along with spread bar and wire grip | Eye and ear protection |
| | Hammer |
| Hog ring pliers | Tape measure |
| 4 × 4 posts | Pencil |
| 2 × 4 lumber | Work gloves |
| | Privacy fabric tape |
| | Vinyl privacy slats |

**Chain link fencing** is a strong, durable, and inexpensive way to create a barrier, increase your property's security, or keep pets safely inside.

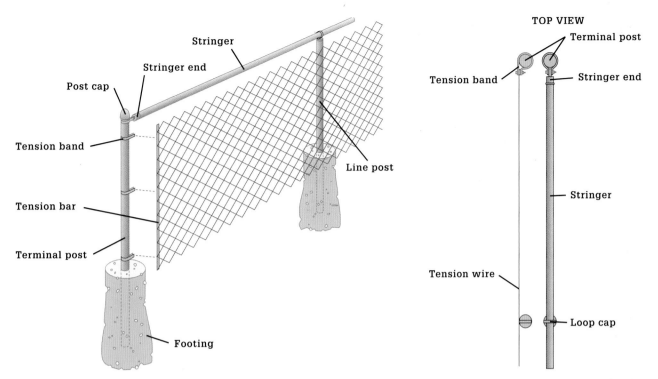

Fittings are designed to accommodate slight alignment and height differences between terminal posts and line posts. Tension bands, which hold the mesh to the terminal posts, have one flat side to keep the mesh flush along the outside of the fence line. The stringer ends hold the top stringer in place and keep it aligned. Loop caps on the line posts position the top stringer to brace the mesh.

## How to Install a Chain Link Fence

**Install the posts.** Lay out the fenceline, spacing the posts at 96" on-center (see pages 22 to 29 for laying out and setting posts). Dig holes for terminal posts 8" in diameter with flared bottoms; dig holes for line posts at 6". Make all postholes 30" deep or below the frost line, whichever is deeper. Set the terminal posts in concrete so they extend 50" above grade. Run a mason's string between terminal posts at 46" above grade. Set the line posts in concrete so their tops are even with the string. If desired, stop the concrete 3" below ground level and backfill with soil and grass to conceal the concrete. *Tip: When plumbing and bracing posts, use duct tape to secure cross bracing to the posts.*

**Position the tension bands** and stringer ends on the gate and end terminal posts, using a ratchet wrench to tighten the bands with the included bolt and nut. Each post gets three tension bands: 8" from the top, 24" from the top, and 8" above the ground (plus a fourth band at the bottom of the post if you will use a tension wire). Make sure the flat side of each band faces the outside of the fence and points into the fence bay. Also add a stringer end to each post, 3" down from the top.

(continued)

**3**

**Add bands and ends to the corner posts.** Each corner post gets six tension bands, two at each location: 8" and 24" from the top and 8" from the bottom (plus two more at the bottom for a tension wire, if applicable). Also install two stringer ends, 3" from the top of the post. Orient the angled side up on the lower stringer end and the angled side down on the upper stringer end.

**4**

**Top each terminal post** with a post cap and each line post with a loop cap. Make sure the loop cap openings are perpendicular to the fenceline, with the offset side facing the outside of the fenceline.

**5**

**Begin installing the stringer,** starting at a terminal post. Feed the non-tapered end of a stringer section through the loop cap on the nearest line post, then into the stringer end on the terminal post. Make sure it's snug in the stringer end cup. Continue feeding stringer sections through loop caps, and join stringer sections together by fitting the non-tapered ends over the tapered ends. If necessary, use a sleeve to join two non-tapered ends.

**6**

**Measure and cut** the last stringer section to fit to complete the stringer installation. Measure from where the taper begins on the preceding section to the end of the stringer end cup. Cut the stringer to length with a hacksaw or pipe cutter. Install the stringer.

**7**

**Secure the chain link mesh to a terminal post,** using a tension bar threaded through the end row of the mesh. Anchor the bar to the tension bands so the mesh extends about 1" above the stringer. The nuts on the tension bands should face inside the fence. If applicable, install a tension wire as directed by the manufacturer. Unroll the mesh to the next terminal post, pulling it taut as you go.

**8**

Spread bar

**Stretch the mesh toward the terminal post** using the come-along. Thread a spread bar through the mesh about 48" from the end, and attach the come-along between the bar and terminal post. Pull the mesh until it's difficult to squeeze the links together by hand. Insert a tension bar through the mesh and secure the bar to the tension bands. Remove excess mesh by unwinding a strand. Tie the mesh to the stringer and line posts every 12" using tie wire. See page 98 to install a gate.

## Weaving Chain Link Mesh Together ▶

If a section of chain link mesh comes up short between the terminal posts, you can add another piece by weaving two sections together.

With the first section laid out along the fenceline, estimate how much more mesh is needed to reach the next terminal post. Overestimate 6" or so, so you don't come up short again.

Detach the amount of mesh needed from the new roll by bending back the knuckle ends of one zig-zag strand in the mesh. Make sure the knuckles of the same strand are undone at the top and bottom of the fence. Spin the strand counter-clockwise to wind it out of the links, separating the mesh into two.

Place this new section of chain link at the short end of the mesh so the zig-zag patterns of the links line up with one another.

Weave the new section of chain link into the other section by reversing the unwinding process. Hook the end of the strand into the first link of the first section. Spin the strand clockwise until it winds into the first link of the second section, and so on. When the strand has connected the two sections, bend both ends back into a knuckle. Attach the chain link mesh to the fence framework.

# How to Install a Chain Link Gate

**Set fence posts in concrete** spaced far enough apart to allow for the width of the gate plus required clearance for the latch. Position the female hinges on the gate frame, as far apart as possible. Secure with nuts and bolts (orient nuts toward the inside of the fence).

**Set the gate on the ground** in the gate opening, next to the gatepost. Mark the positions of the female hinges onto the gate post. Remove the gate and measure up 2" from each hinge mark on the gatepost. Make new reference marks for the male hinges.

**Secure the bottom male hinge** to the gatepost with nuts and bolts. Slide the gate onto the bottom hinge. Then, lock the gate in with the downward-facing top hinge.

**Test the swing of the gate** and adjust the hinge locations and orientations, if needed, until the gate operates smoothly and the opposite side of the gate frame is parallel to the other fence post. Tighten the hinge nuts securely.

**Attach the gate latch to the free side of the gate frame,** near the top of the frame. Test to make sure the latch and gate function correctly. If you need to relocate a post because the opening is too large or too small, choose the latch post, not the gate post.

# How to Build a California-style Chain Link Fence

**Install the posts.** Set the 4 × 4 fence posts in concrete, spacing them at 6 to 8 ft. on center. The posts should stand at least 4" taller than the finished height of the chain link mesh. See pages 22 to 29 for help with laying out your fenceline and installing the posts.

**Trim the posts** so they are 4" higher than the installed height of the chain link mesh. Mark the post height on all four sides of each post, and make the cuts with a circular saw, reciprocating saw, or handsaw.

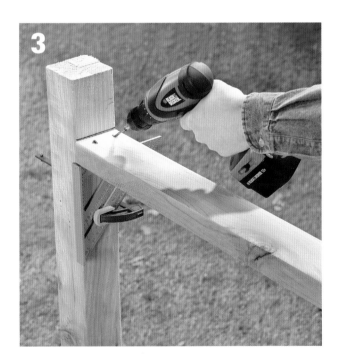

**Add 2 × 4 top stringers** between each pair of posts. Mark reference lines 4" down from the tops of the posts. Cut each stringer to fit snugly between the posts. Fasten the stringers with their top faces on the lines using 3" deck screws or 16d galvanized common nails driven through angled pilot holes.

**Wrap tension wire around a terminal post,** about 1" above the ground. Staple the wire with a galvanized fence staple, and then double back the tail of the wire and staple it to the post.

(continued)

**Staple the tension wire** to the line posts after gently tightening the wire (using a come-along with a wire grip) and securing the loose end of the wire to the opposing terminal post. Option: You can install 2 × 4 bottom stringers in place of a tension wire.

**Add finials or decorative caps** to the post tops for a finished look and to help protect the end grain of the wood.

**Secure the fence mesh** to the first terminal post using a tension bar threaded through the end row of the mesh. Fasten the bar to the posts with a fence staple every 8". Make sure the bar is plumb and the top of the mesh overlaps the top stringer (and bottom stringer, if applicable).

**Unroll the mesh toward the other terminal post,** and then stretch the mesh gently with a come-along (see step 8, page 97). Secure the end of the mesh to the post with a tension bar and staples, as before. Remove any excess mesh by unwinding a strand (see page 97).

**Attach the bottom edge of the mesh** to the tension wire every 2 ft., using hog rings tightened with hog ring pliers. Staple the mesh to the stringers every 2 ft. and to the line posts every 12".

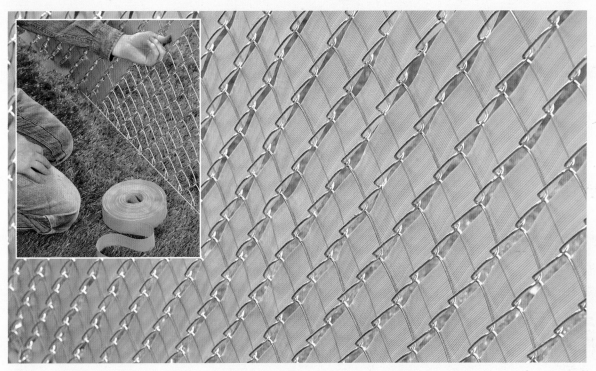

**Privacy fabric tape** cuts the wind and provides partial privacy. It's purchased in rolls with a limited number of color options. It is relatively inexpensive, but threading it through the chain link mesh is time consuming.

**Vinyl privacy slats** create vertical lines and are easier to install than tape. They're available in a limited number of colors at most building centers. Some varieties of strips also have a grass-like texture.

# Trellis Fence

This simple design creates a sophisticated trellis fence that would work in many settings. Part of its appeal is that the materials are inexpensive and the construction remarkably simple.

It can be used as a accent feature, a backdrop to a shallow garden bed, or as a screen to block a particular view. As a vertical showcase for foliage or flowers, it can support a wide display of colorful choices. Try perennial vines such as Golden Clematis or Trumpet Creeper. Or, for spectacular autumn color, plant Boston Ivy. If you prefer annual vines, you might choose Morning Glories or a Black-eyed Susan Vine. The possibilities go on and on—just make sure that the plants you select are well-suited to the amount of sunlight they'll receive.

Depending on the overall look you want to achieve, you can paint, stain, or seal the fence to contrast with or complement your house or other established structures. Well-chosen post finials can also help tie the fence into the look of your landscape.

This project creates three panels. If you adapt it to use a different number of panels, revise the materials list accordingly.

**A trellis fence** is a decorative way to incorporate minimal privacy. The basic installation techniques shown in this project will allow you to put your trellis to creative use. Shown here, the trellis ties into an arbor gate—a beautiful transition from yard to garden.

# Tools, Materials & Cutting List

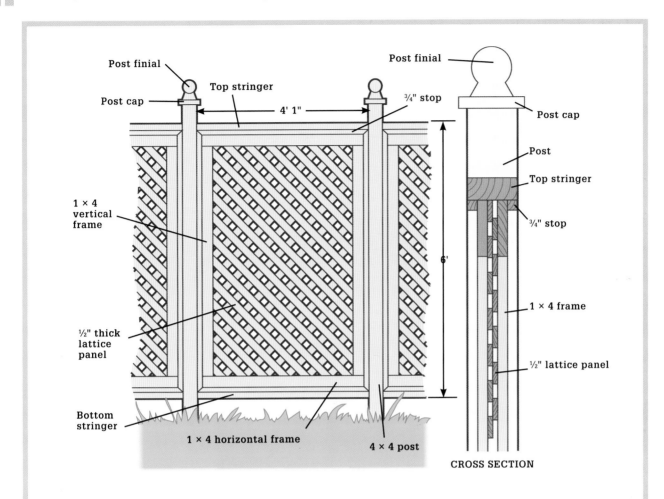

Post finial

Post cap

Top stringer

4' 1"

¾" stop

Post finial

Post cap

Post

Top stringer

¾" stop

1 × 4 vertical frame

6'

1 × 4 frame

½" thick lattice panel

½" lattice panel

Bottom stringer

1 × 4 horizontal frame

4 × 4 post

**CROSS SECTION**

Tools and materials for setting posts
Mason's line
Line level
Dowel screws
Hammer
Tape measure
Circular saw or reciprocating saw
Drill
Caulk gun
Nail set
Pressure-treated cedar or redwood lumber
   (see cutting list)
½" lattice panels (3)
10d corrosion-resistant casing nails
Corrosion-resistant finish nails (4d, 6d)
Construction adhesive
Deck post finials (4)
Framing square or straightedge
Eye and ear protection
Work gloves

| Part | Lumber | Size | Number |
|------|--------|------|--------|
| Posts | 4 × 4 | 10 ft. | 4 |
| Stringers | 2 × 4 | cut to fit | 6 |
| Back frame | | | |
| Top & bottom | 1 × 4 | 41½" | 6 |
| Sides | 1 × 4 | 71¾" | 6 |
| Front frame | | | |
| Top & bottom | 1 × 4 | 48½" | 6 |
| Sides | 1 × 4 | 64¾" | 6 |
| Stops | | | |
| Top & bottom | 1 × 1 | cut to fit | 12 |
| Sides | 1 × 1 | cut to fit | 12 |
| Lattice panels | 4 × 8 | 48½ × 71¾" | 3 |
| Post caps | 1 × 6 | 4½ × 4½" | 4 |

# How to Build a Trellis Fence

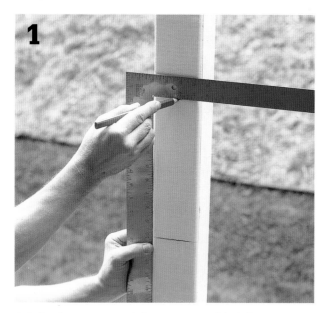

**1**

**Set the four 4 × 4 posts in concrete,** with their tops at least 84" above the ground, and space them 52½" on-center (see pages 22 to 29 for help with laying out postholes and setting posts). On one of the end posts, make a mark at 77" and 83" above the ground. Using a level mason's line, transfer both height marks to all of the posts, and then mark all sides of each post with a square. Trim the posts at the 83" mark with a circular saw or reciprocating saw.

**2**

**Install the stringers.** On each post, make a mark 72" down from the 77" mark made in step one. Measure and cut each 2 × 4 stringer to fit between these respective marks on each post pair. Install the top stringer on top of the 77" mark, and install the bottom stringer on the bottom of the lower mark. The framed opening should measure 49" × 72". Fasten the stringers to the posts with 10d corrosion-resistant casing nails driven through angled pilot holes.

**3**

**Add the stops** to the back edges of the fence frames. Measure and cut the 1 × 1 stops to fit the framed openings. You can miter the stop pieces or fit them together with simple butt joints. Position the stops so they are flush with the back sides of the posts and stringers, and fasten them to the posts and stringers with 6d corrosion-resistant finish nails driven through pilot holes.

**4**

**Prepare the lattice frames.** Cut all of the 1 × 4 frame pieces, following the dimensions in the cutting list on page 103. Assemble each back frame by butting the pieces together, and measuring diagonally between opposing corners to make sure the frame is square (frame is square when the diagonal measurements are equal). Cut the lattice panels at 48½" × 71¾". Apply a wavy bead of construction adhesive along the center of the back frame boards.

**Complete the lattice frames.** Set the lattice panel into the adhesive on each back frame so the panel edges are flush with the frame boards. Position the front frame pieces over the lattice (the butted joints in the front frame should be offset from those in the back frame). Fasten the frame boards and lattice together with 4d finish nails.

**Install the lattice frames.** Measure and cut each set of front stops to fit a fence opening. Set each lattice frame into an opening and hold it against the back stops. Fit the front stops over the lattice frame one at a time, pressing tightly against the frame. Fasten the stops to the posts with 6d finish nails.

**Add the post caps and finials.** Cut the post caps from 1 × 6 lumber, making them 4½" square. Center each cap over a post and secure it with 6d finish nails. Mark the center of each cap by drawing an × between opposing corners using a square or straightedge. At the centerpoint, drill a pilot hole for a dowel screw (a screw with coarse threads on both ends). Secure the finial to the cap with a dowel screw.

## Easy Plant Ties ▸

Tying vines requires a material that's both strong and gentle—strong enough to support the vine and gentle enough not to damage the tendrils.

Old 100 percent cotton t-shirts make terrific, inexpensive ties that can go into the compost bin for further recycling when the growing season is over.

Starting at the bottom, cut around the shirt in a continuous spiral about 1½" wide. When you reach the armholes, begin making straight cuts from the edge of one sleeve to the edge of the other. One shirt usually produces 15 to 20 yards of tying material.

# Bamboo Fence

Bamboo is one of nature's best building materials. It's lightweight, naturally rot-resistant, and so strong that it's used for scaffolding in many parts of the world. It's also a highly sustainable resource, since many species can be harvested every three to five years without destroying the plants. Yet, perhaps the best feature of bamboo is its appearance—whether it's lined up in orderly rows or hand-tied into decorative patterns, bamboo fencing has an exotic, organic quality that adds a breath of life to any setting.

Bamboo is a grass, but it shares many properties with wood. It can be cut, drilled, and sanded with the same tools, and it takes many of the same finishes, including stains and exterior sealers. And, just like wood, bamboo is prone to splitting, though it retains much of its strength even when subject to large splits and cracks. In general, larger-diameter poles (which can be upwards of 5 inches) are more likely to split than smaller (such as ¾-inch-dia.) canes.

Bamboo fencing is commonly available in eight-foot-long panels made from similarly sized canes held together with internal or external wires. The panels, which are rolled up for easy transport, can be used as infill within a new wood framework, or they can attach directly to an existing wood or metal fence. Both of these popular applications are shown here. Another option is to build an all-bamboo fence using large bamboo poles for the posts and stringers and roll-up panels for the infill.

## Tools & Materials ▸

| | |
|---|---|
| Tools and materials for laying out and setting posts | 1 × 4, 2 × 6) |
| Circular saw or reciprocating saw | Deck screws (3", 2½", 2") |
| Drill | Bamboo fence panels with ¾"-dia. (or as desired) canes |
| Countersink-piloting bit | Level |
| Wire cutters | Tape measure |
| Pliers | Eye and ear protection |
| Lumber (4 × 4, 2 × 4, | Galvanized steel wire |
| | Work gloves |

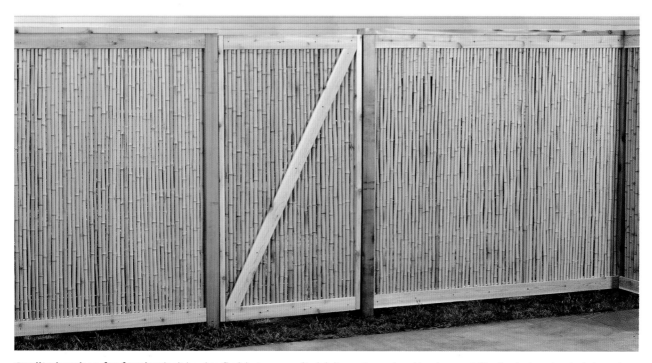

**Quality bamboo for fencing** isn't hard to find, but you can't pick it up at your local lumberyard. The best place to start shopping is the internet (see Resources, page 204). Look for well-established suppliers who are committed to sustainable practices. Most suppliers can ship product directly to your home.

# How to Build a Wood-frame Bamboo Fence

**Install and trim** the 4 × 4 posts according to the size of your bamboo panels, setting the posts in concrete (see pages 22 to 29). For the 6 × 8-ft. panels in this project, the posts are spaced 100" on-center and are trimmed at 75" tall (refer to the manufacturer's recommendations).

**Install the top 2 × 4 stringers.** Cut each stringer to fit snugly between the posts. Position the stringer on edge so it is flush with the tops of the posts and with the back or front faces of the posts. Fasten the stringer with 3" deck screws driven through angled pilot holes. Use one screw on each edge and one on the inside face of the stringer, at both ends.

**Mark the location of each bottom stringer.** The span between the top of the top stringer and bottom of the bottom stringer should equal the bamboo panel height plus about 1". Cut and install the bottom stringers in the same fashion as the top stringers. Here, the bottom stringer will be installed 2" above the ground for rot prevention. Unroll the bamboo panels.

**Flatten the bamboo panels** over the inside faces of the stringers. Make sure the panels fit the frames on all sides. Using a countersink-piloting bit (inset), drill a slightly countersunk pilot hole through a bamboo cane and into the stringer at a top corner of the panel. Fasten the corner with a 2" deck screw, being careful not to overtighten and split the bamboo.

(continued)

**5**

Screws       2 × 4

**Fasten the rest of the panel with screws** spaced 12" apart. Stagger the screws top and bottom, and drive them in an alternating pattern, working from one side to the other. Repeat steps 4 and 5 to install the remaining bamboo panels.

## Reducing Panel Length ▶

**To shorten the length of a bamboo panel,** cut the wiring holding the canes together at least two canes beyond the desired length using wire cutters. Remove the extra canes, and then wrap the loose ends of wire around the last cane in the panel.

**6**

**Cover the top and bottom ends** of the panels with 1 × 4 battens. These finish off the panels and give the fence a similar look on both sides. Cut the battens so the ends are flush against the inside faces of the posts and fasten them to the panels and stringers with 2½" deck screws driven through pilot holes.

**7**

**Add the top cap.** Center the 2 × 6 top cap boards over the posts so they overhang about 1" on either side. Fasten the caps to the posts and stringers with 3" deck screws. Use miter joints for corners, and use scarf joints (cut with opposing 30° or 45° bevels) to join cap boards over long runs.

# How to Cover an Old Fence with Bamboo

**Unroll and position a bamboo panel** over one or both sides of the existing fence. Check the panel with a level and adjust as needed. For rot prevention, hold the panel 1 to 2" above the ground. *Tip: A 2 × 4 laid flat on the ground makes it easy to prop up and level the panel.*

**Fasten the panel with deck screws** driven through the bamboo canes (and fence siding boards, if applicable) and into the fence stringers. Drill countersunk pilot holes for the screws, being careful not to overtighten and crack the bamboo. Space the screws 12" apart, and stagger them top and bottom (see pages 107 and 108).

**Install the remaining bamboo panels,** butting the edges together between panels for a seamless appearance. If the fence posts project above the stringer boards, you can cut the bamboo panels flush with the posts. To trim the panels, follow the technique shown in the Tip on page 108.

**Variation:** To dress up a chain link fence with bamboo fencing, simply unroll the panels over the fence and secure them every 12" or so with short lengths of galvanized steel wire. Tie the wire around the canes or the panel wiring and over the chain link mesh.

# Invisible Dog Fence

The invisible, or underground, pet fence can be the perfect option for those who love dogs but not necessarily dog fences. The pet fence is invisible because the actual boundary is nothing more than a thin electrical wire buried an inch or so underground. It can also be laid into hard walkway and driveway surfaces and can be installed above ground, on fences and other fixed structures. This makes it easy to create a continuous barrier to enclose any or all of your property as well as specific areas inside the boundary, such as a garden or swimming pool. Invisible fences can also be used for cats, provided they meet the weight minimum for safe use.

Here's how the pet fence works: the boundary wire receives a constant electrical signal from a small, plug-in transmitter located in the house or other protected space. Your pet is fitted with a special collar that picks up the signal in the wire and responds accordingly: if your pet approaches the boundary area, the collar beeps and vibrates to give him warning that the boundary is near. If he continues beyond the warning zone, he is given an electrical shock by the collar contacts—a clear message to back away from the boundary.

Invisible pet fences effectively contain dogs of all types and are suitable for small and large properties—up to 10 acres, in some cases. However, it's important to make sure this type of system is right for your needs and your pet. While most dogs quickly learn to respect the system, it's always possible for a dog to breach the boundary (a high-spirited pooch may be especially prone to doing so). And keep in mind that this fence will not prevent other dogs from entering your yard. After installing the fence, it's critical that you take the time to train your dog properly so that he knows where the boundary lines are and understands the correction system. The fence manufacturer should provide detailed training instructions.

## Tools & Materials ›

| | |
|---|---|
| Tape measure (100-ft.) | Screws |
| Drill | Stapler and staples |
| Straightened coat hanger | (for wood fence |
| Flat spade | installation only) |
| Paint stir-stick | Electrical tape |
| Circular saw and | Zip ties (for |
| masonry blade | metal fence |
| Concrete or | installation only) |
| asphalt caulk or | Wire stripper |
| patching material | Wire nuts |
| Shop vacuum | Silicone caulk |
| Pet fence kit | Caulk gun |
| Eye and ear potection | Work gloves |

**A properly trained pet** will stay within the invisible boundary, as long as the animal wears the collar that is part of the invisible fence system.

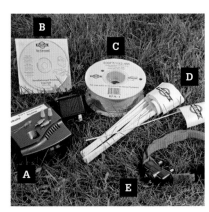

**Invisible pet fence systems** are available in complete kits and can be installed in a day. The basic components for installation (photo above) include from left to right: a transmitter and power cord (A), installation manual or disc (B), boundary wire (C), boundary flags (D), and receiver collar (E).

# Boundary Layout Options

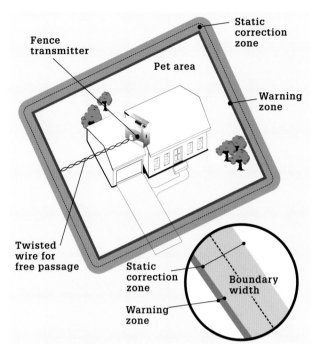

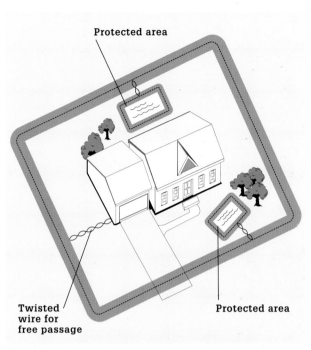

**A perimeter layout** uses a single run of wire encircling the house and grounds. A single section of twisted wire runs from the boundary to the transmitter. *Note: Twisting the boundary wire around itself cancels the signal, creating a "free passage" area for your pet.*

**Protecting areas within a perimeter boundary** is achieved by looping the wire around the area and returning to the boundary. Twisting the wire between the boundary and protected inner area allow for free passage around the protected area.

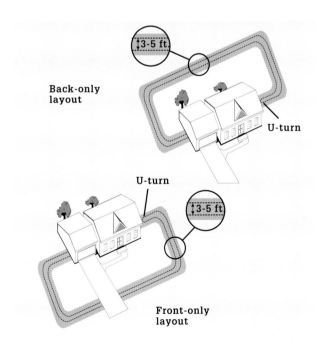

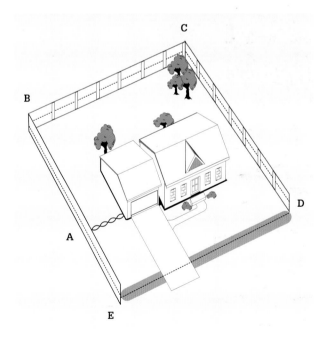

**A front or back-only layout** requires a doubled loop of wire to complete the boundary circuit. Starting at the transmitter, the wire encircles the containment area and then doubles back, maintaining a 3 to 5-ft. space (or as directed) between runs to prevent canceling the signal.

**Incorporating a fence into the boundary** can help deter your dog from jumping over or digging under the fence. The wire can be fastened directly to the fence and/or can be buried in front of the fence. Burial allows you to protect gate openings. Run wires from the transmitter to A, A to B, B to C, C to D, D to E, E to A, and then twist wire from A to transmitter.

# How to Install an Invisible Dog Fence

**Plan the layout of the boundary wire.** With a helper, use a 100-ft. tape measure to determine the total distance of the wire run. Factor in extra length for twisted (free passage) sections and for making adjustments. Order additional wiring, if necessary. *Tip: Use the boundary flags that come with the kit to temporarily mark the corners and other points of the wire route.*

**Mount the transmitter** on the inside of an exterior wall, near a standard 120-volt receptacle. The location can be in the house, garage, basement, or crawlspace and must be convenient, protected from the elements, not subject to freezing temperatures, and must be at least 3 ft. (or as recommended) from appliances or other large metal objects. Mount the transmitter with appropriate screws.

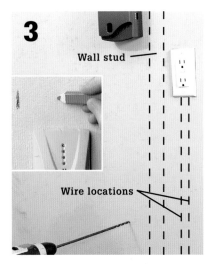

Wall stud

Wire locations

**Drill a hole through the wall** for routing the boundary wire. The hole can be just large enough to fit the wire (which will likely be twisted at this point; see step 5, page 113). Alternatively, you can route the wire through a window, door, or crawlspace/basement-wall vent, provided the wire will be safe from damage. Identify stud and wire locations before you drill but shut off electrical power to be safe.

**Begin running the wire along the planned route.** Be sure to leave extra wire for twisting at the termination point of the boundary (transmitter location), if applicable. Turn corners with the wire gradually, not at sharp angles.

**5**

**Twist the wire onto itself** to cancel the signal for free passage areas, as desired. With a helper holding the wire at the end of a loop (start of twisted section), circle the spool around the wire to create 10 to 12 twists per linear foot (or as recommended). Be sure not to exceed the maximum length of twisted wire.

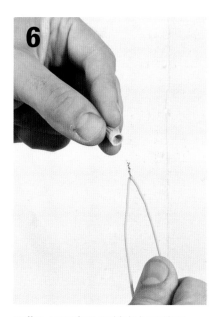

**6**

**Splice together** multiple boundary wires (required only when the boundary distance exceeds the length of wire provided with kit). Strip ½" of insulation from the ends of both wires using a wire stripper. Hold the ends together and join them with a wire nut, twisting the nut on tightly. Tug on the wires to make sure they're held by the nut.

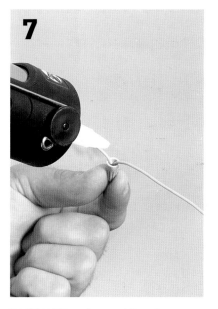

**7**

**Seal inside** and around the wire nut with silicone caulk to create a waterproof connection and prevent corrosion. When the caulk has dried completely, reinforce the connection with electrical tape. *Tip: Make note of each splice location, as these are the most common points of boundary wire failure.*

**8**

**Fish the ends of the boundary wire** through the house wall (termination of the boundary run) using a straightened coat hanger. Wrap the wire ends around the fish tape bend and secure them with electrical tape. Once through the wall, twist the wires to provide free passage from the house to the boundary line, as appropriate for your layout.

(continued)

**Connect the wire ends** to the transmitter after stripping ½" of insulation from each end. Secure the wires to the appropriate terminals on the transmitter. Plug in the transmitter and set the boundary controls for testing the system, as directed.

**Test the system** using the receiver collar and testing tool. Adjust the collar settings as directed. Walk toward the boundary wire while holding the collar at the pet's neck height. Note when the warning and correction signals are activated, indicated by the testing tool. Test the system in multiple locations. Make adjustments to the settings and/or boundary wire as needed.

**Excavate grass or soil,** making a continuous cut 1 to 3" deep. Drive a spade into the ground, then rock the handle back and forth to widen the cut slightly, creating a straight or gently curving slot.

**Lay the boundary wire into the slot,** using a paint stir-stick to seat it into the bottom of the slot. Be careful not to kink or damage the wire. Carefully close the slot by stepping along its length with one foot on either side of the slot.

**Cut slots into concrete** or asphalt driveways and walks using a circular saw with a masonry blade. Vacuum the slot clean, and then lay the boundary wire into the slot. Seal over the slot with high-quality concrete or asphalt caulk or patching compound.

**Option:** Use existing control joints to pass the boundary wire over concrete drives and walks. Control joints are the shallow grooves formed in the concrete to help control cracking. Clean and vacuum the joint, then lay in the wire. Cover the joint with concrete caulk.

**Fasten the boundary wire to fences,** as directed by the manufacturer. Use staples for wood fences and plastic zip ties for metal fences (or simply weave the wire through chain link mesh). To protect gate openings, bury the wire in the ground in front of the opening.

**Position the boundary flags** using the collar to find the inside edge of the warning zone. Move toward the boundary until the collar beeps (warning signal) and place a flag at that location. Place a flag every 10 ft. (or as directed) over the entire boundary area. Fit the collar to your pet as directed to begin the training. After training period, remove flags (follow manufacturer instructions).

# Brick & Cedar Fence

This elegant fence is an attractive, durable structure that will be the envy of the neighborhood. The 72-inch-tall brick pillars can be used in place of posts with wood or ornamental metal fencing. If you'd like to use metal fence panels instead of the board and stringer wood panels built in this project, consult the fencing manufacturer about options for anchoring the panels to the brick pillars. Make sure to have the panels and hardware on hand when laying out and building the pillars, to make sure the pillar spacing is accurate.

Each pillar must be built on its own concrete footing that extends below the frost line; in this project, the footings are 16 inches wide × 20 inches long. See pages 40 to 45 for tips on working with concrete and instructions for building structural footings. To maintain an even ⅜" mortar joint spacing between bricks, create a story pole using a 2 × 2 marked with the spacing. After every few courses, hold the pole against the pillar to check the joints for a consistent thickness. Also make sure the pillars remain as plumb, level, and square as possible to increase the strength and longevity of the pillars.

Attaching the stringers to the pillars is much easier than you may imagine. Fence brackets and concrete screws are available that have as much holding power as lag bolts and anchors. Although other brands are available, we used ¼"-dia. TapCon concrete screws. The screws come with a special drill bit to make sure the pilot holes are the right diameter and depth, which simplifies the process for you.

## Tools & Materials ▸

Tools and materials for setting footings
Level
Wheelbarrow or mixing box
Mason's trowel
Jointing tool
Aviation snips
Drill
Circular saw
Jigsaw
Standard modular bricks (4 × 2⅔ × 8", 130 per pillar)
2 × 2 lumber for story pole
Type N mortar mix
¼" wooden dowel
¼" wire mesh
Capstone or concrete cap
⅜"-thick wood scraps
1¼" countersink concrete screws

2 × 6 corrosion-resistant fence brackets (6 per bay)
Concrete drill bit
Pressure-treated cedar or redwood lumber:
1 × 6, 12 ft. (8 per bay)
2 × 6, 8 ft. (3 per bay)
Corrosion-resistant deck screws (1¼", 1½")
1½" finish nails
96"-length of flexible ¼" PVC pipe
Vegetable oil
Chalk
Chalk line
Tape measure
Eye and ear protection
Work gloves

**This elegant fence** creates a stable boundary, and adds aesthetically pleasing textures to your home landscape.

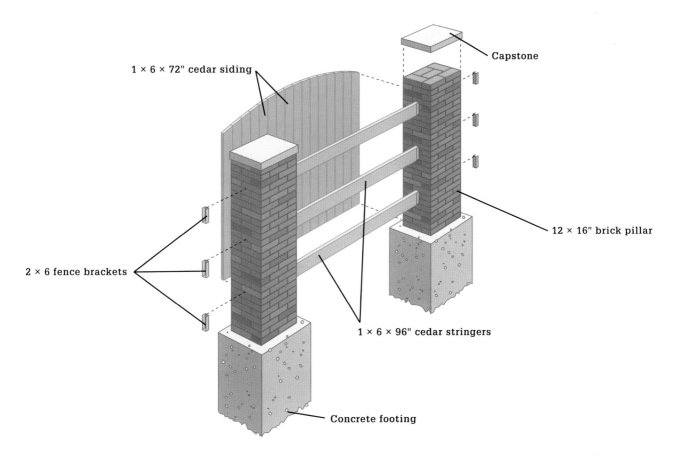

Capstone

1 × 6 × 72" cedar siding

12 × 16" brick pillar

2 × 6 fence brackets

1 × 6 × 96" cedar stringers

Concrete footing

# How to Build a Brick & Cedar Fence

**Mark the outline of the first course of brick** on the cured concrete footing. Dry-lay the five bricks of the course so they are centered on the footing. Leave a ⅜" gap for each mortar joint. Remove the bricks, apply a bed of mortar inside the reference lines, and then press the first course of brick into the mortar to create a ⅜"-thick bed layer.

**Use a pencil or dowel coated with vegetable oil** to create a weep hole in the mortar in the first course of bricks before filling in the joints. The hole ensures drainage of any moisture that seeps into the pillar. Fill in the joint and then remove dowel.

(continued)

**Lay the second course,** rotating the pattern 180°. Lay additional courses, rotating the pattern 180° with each course. Use the story pole and a level to check each face of the pillar after every other course. (It's important to check frequently, since any errors will be exaggerated with each successive course.)

**After every fourth course,** cut a strip of ¼" wire mesh and place it over a thin bed of mortar. Add another thin bed of mortar on top of the mesh, then add the next course of brick.

**After every five courses,** use a jointing tool to smooth the joints that have hardened enough to resist minimal finger pressure.

**For the final course,** lay the bricks over a bed of mortar and wire mesh. After placing the first two bricks, add an extra brick in the center of the course. Lay the remainder of the bricks around it. Fill the remaining joints, and work them with the jointing tool as soon as they become firm.

**Lay the concrete cap stone.** Mark reference lines on the underside of the cap to help you center it on the pillar. Spread a ½"-thick bed of mortar on top of the pillar. Center the cap on the pillar, and then strike the mortar joint under the cap so it's flush with the brick. *Note: If mortar squeezes out of the joint, press ⅜"-thick wood scraps into the mortar at each corner to support the cap. Remove the scraps after 24 hours and fill in the gaps with mortar.* Let the pillars cure, as directed.

**Install the fence brackets and stringers.** Mark the inside face of each pillar at 18, 36, and 60" down from the top using chalk. At each mark, make a second mark 6¾" in from the outside face of the pillar. Position a 2 × 6 fence bracket at each mark intersection, and fasten the bracket to the pillar with concrete screws driven through pilot holes (make the holes ¼" deeper than the screw shaft). Cut a 2 × 6 stringer to fit between each corresponding pair of brackets. Attach the stringers with 1¼" deck screws.

**Mark and cut the fence boards.** Cut the 12-ft. 1 × 6 boards in half. Set 16 pieces together on a flat surface, with their bottom ends aligned, setting a ½" (approx.) gap between each board. On each end board, drive a finish nail 64" up from the bottom and 2" in from the outside edge. Snap a chalk line between the nails. At the center of the middle board, drive a third nail 70" from the bottom. Bend flexible PVC pipe by placing it under the side nails and over the center nail. Trace along the pipe to mark the arch. Cut out the arch with a jigsaw and remove nails.

**Install the fence boards.** Starting at one pillar, position the first board so it is plumb and 2" above the ground. Fasten it to the stringers with pairs of 1½" deck screws driven through pilot holes. Install the remaining boards, using the chalk line to align the boards for proper height. Maintain a ½"-gap between boards, and check every few boards with a level to make sure they're plumb.

# Stone & Rail Fence

This 36"-tall, rustic stone and rail fence is constructed in much the same way as the brick and cedar fence, but with stone rather than brick and simple 2 × 4 rails rather than siding.

Each pillar requires a footing that extends 6" beyond its base in all directions. Carefully plan the layout and sort the stones before you begin setting the stone. If necessary, use a stone cutter's chisel and a maul to trim stones or cut them to size.

## Tools & Materials ▸

Tape measure
Level
Wheelbarrow
    or mixing box
Mason's trowel
Jointing tool
Stone cutter's chisel
    & maul
Stiff-bristle brush
Drill
Paintbrush and roller
Type M mortar
Wood shims
Stones of various
    shapes and sizes
Fence brackets
    (2 × 4; 6 per bay)

1¼" countersink
    concrete screws
Concrete drill bit
Rough-cut cedar
    2 × 4s, 8 ft.
    (3 per bay)
Paint, stain, or sealer
1½" corrosion-resistant
    deck screws
Stakes
Mason's string
Shovel
Eye and ear protection
Chalk
Work gloves
Circular saw or
    reciprocating saw

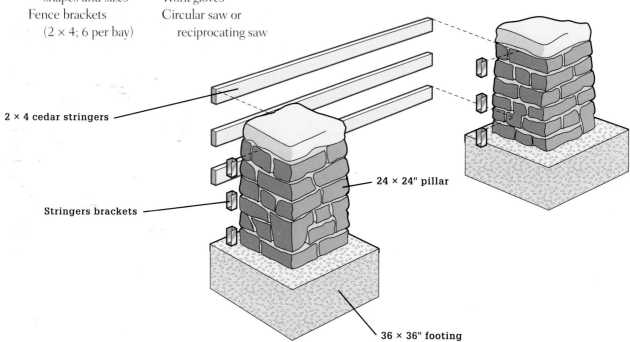

2 × 4 cedar stringers

Stringers brackets

24 × 24" pillar

36 × 36" footing

# How to Build a Stone & Rail Fence

### DRY-LAY THE FIRST COURSE

Plot the fenceline with stakes and mason's string (pages 22 to 25). For 72" bays between 24 × 24" pillars, measure and mark 18" in from the end of the fenceline and then every 96" on-center. Outline, dig, and pour 36 × 36" concrete footings. Let the concrete cure for two days.

Sort individual stones by size and shape. Set aside suitable tie stones for corners and the largest stones for the base. Dry-lay the outside stones in the first course to form a 24 × 24" base centered on the footing. Use chalk to trace a reference outline on the footing around the stones, then set them aside.

### MORTAR THE FIRST COURSE

Trowel a 1"-thick bed of mortar inside the reference outline, then place the stones in the mortar, in the same positions as in the dry-run. Fill in the center with small stones and mortar (photo 1). Leave the center slightly lower than the outer stones. Pack mortar between the outer stones, recessing it roughly 1".

### LAY MORE COURSES & TOOL THE JOINTS

Set each subsequent course of stone in a bed of mortar laid over the preceding course, staggering the vertical joints (photo 2). On every other course, place tie stones that extend into the pillar center. Use wood shims to support large stones until the mortar sets. Build each pillar 36" tall using a level to check for plumb. When the mortar sets enough to resist light finger pressure, smooth the joints with a jointing tool. Remove any shims and fill the holes with mortar. Remove dry spattered mortar with a stiff-bristle brush.

### LAY TOP CAP & ATTACH THE STRINGERS

Lay a 1"-thick bed of mortar on the pillar top and place the capstones (photo 3). Smooth the joints. Mist with water regularly for one week, as the mortar cures.

On the inner face of each pillar, measure up from the footing and mark with chalk at 12, 21, and 30". At each mark, measure in 6" from the outside face of the pillar and mark, then line up a 2 × 4 fence bracket where these two marks intersect. Mark the screw holes on the pillar, then drill a 1½"-deep embedment hole at each mark.

Align the bracket screw holes with the embedment holes, and attach with 1¼" countersink concrete screws. Repeat for each bracket. Measure the distance from a fence bracket on one pillar to the corresponding bracket on the next for the stringer size. Mark and cut stringers to size. Insert stringers into the fence brackets and attach them with 1½" corrosion-resistant screws.

**Spread a 1"-thick bed of mortar** on top of the footing, and beginning stacking the stones inside the pillar outline. Fill gaps between stones with mortar and spread a bed of mortar over the first-course stones.

**Build up courses of stones,** checking with a level and adding or subtracting mortar as you go to keep each course level. Fill gaps between stones and smooth with a jointing tool.

**Spread a 1"-thick bed of mortar** on the top course of stones and set the cap block into the mortar. Clean off excess mortar and smooth out the mortar joints with a jointing tool. Attach fence rail hangers and fence rails as described.

# Easy Custom Gates

If you understand the basic elements of gate construction, you can build a sturdy gate to suit almost any situation. The gates shown here illustrate the fundamental elements of a well-built gate.

To begin with, adequate distribution of the gate's weight is critical to its operation. Because the posts bear most of a gate's weight, they're set at least 12 inches deeper than fence posts. Or, depending on building codes in your area, they may need to be set below the frost line in substantial concrete footings.

However they're set, the posts must be plumb. A sagging post can be reinforced by attaching a sag rod at the top of the post and running it diagonally to the lower end of the next post. Tighten the knuckle in the middle until the post is properly aligned. A caster can be used with heavy gates over smooth surfaces to assist with the weight load.

The frame also plays an important part in properly distributing the gate's weight. The two basic gate frames featured here are the foundation for many gate designs. A Z-frame gate is ideal for a light, simple gate. This frame consists of a pair of horizontal braces with a diagonal brace running between them. A perimeter-frame gate is necessary for a heavier or more elaborate gate. It employs a solid, four-cornered frame with a diagonal brace attached at opposite corners.

In both styles, the diagonal brace must run from the bottom of the hinge-side to the top of the latch-side to provide support and keep the gate square.

Buy your gate hardware before you build your gate, since it will affect the clearance between gate and post. Take the diameter of your gatepost and a drawing of your gate to the store to purchase the correct hardware. The placement and orientation of your gate framing and the width of your posts will affect which hinges and latches may be properly secured through the siding and into the gateposts and gate framing.

## Tools & Materials ▸

| | |
|---|---|
| Tape measure | Pressure-treated cedar |
| Level | or redwood lumber |
| Framing square | as needed: |
| Circular saw | 1 × 2s, 2 × 4s |
| Drill | Gate handle or latch |
| Spring clamps | Galvanized deck |
| Hinge hardware | screws (2", 2½") |
| Paintbrush | Work gloves |
| Paint, stain, or sealer | Eye and ear protection |

**A sturdy gate** provides attractive access to a garden, path, or to your front door. Match your custom gate to the fence for a polished look.

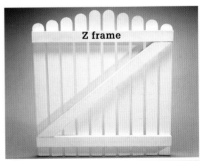

Z frame

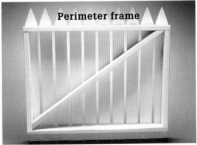

Perimeter frame

**A Z-frame gate** is ideal for lightweight materials. More elaborate or heavy-duty gates will require a perimeter frame for additional stability.

# How to Build a Z-frame Gate

**Check both posts** for plumb and measure the gate opening. Consult your hinge and latch hardware for necessary clearances, and subtract that amount from the opening width. Cut 2 × 4s to this length. Paint, stain or seal the lumber for the gate and let it dry.

**Measure the distance** between the top and bottom stringers on the fence. Cut two 2 × 4s to this length to use as supports. Lay out the frame with the supports between the braces. Square the corners with a framing square. Place a 2 × 4 diagonally across the frame and mark and cut it to length. Screw the brace into position with 2½" deck screws.

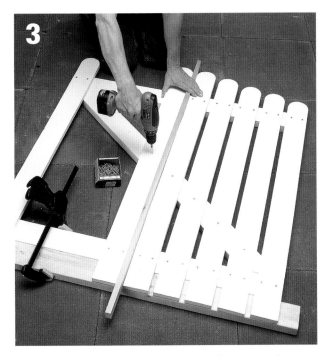

**Plan the layout of the siding to match the fence.** Clamp a 2 × 4 against the bottom brace. Align the first and last boards with the 2 × 4 and attach them to the frame using pairs of 2" deck screws. Attach the rest of the siding using spacers as necessary.

**Shim the gate into position** and mark the locations for the hinges. Drill pilot holes and install the hinges. Hang the gate with the hardware provided and install the latch hardware according to the manufacturer's instructions.

# How to Build a Perimeter-frame Gate

**Follow step 1 for a Z-frame gate** (page 123). Measure the distance between the top and bottom stringers on the fence. Cut two 2 × 4s to this length for the vertical braces. Lay out the frame and secure with 2½" deck screws.

**Position the frame** on a 2 × 4 set on edge running diagonally from one corner to the other. Use scrap 2 × 4 to support the frame. Mark the diagonal brace and cut to length using a circular saw set to the desired bevel angle. Screw the brace into position with 2½" deck screws.

**Clamp a scrap 2 × 4** to the bottom edge to use as a guide. Align the first and last siding boards flush with the edges of the vertical braces. Attach the boards with 2" deck screws. Align and attach the rest of the siding using a spacer. Mount the hardware and hang the gate as for a Z-frame gate (page 123).

**Variation:** Dress up your gate with inset accent pieces. This gate has a frame with a horizontal support to allow the inset of a stained glass window.

# Prefabricated Gate Options

**Chain link gates** are available in a limited range of sizes and styles at home centers. Specialty retailers may offer more options.

**Iron gates** are available as prefab units or can be custom-built for your needs. The strength and beauty of iron pair well with stone or brick walls and entry columns.

**Painted aluminum gates** are a lightweight option that can blend well with any fence style.

**Vinyl fencing manufacturers** offer a suprisingly wide selection of custom gates to help you individualize your fence and landscape.

# Arched Gate

With its height and strategically placed opening, this gate is a great choice for maintaining privacy and enhancing security with style. No ordinary "peephole," the decorative wrought iron provides a stunning accent and gives you the opportunity to see who's heading your way or passing by. The arch of the gate also adds contrast to the fenceline and draws attention to the entryway.

This gate is best suited to a situation where you can position it over a hard surface, such as a sidewalk or driveway. The combined weight of the lumber and the wrought iron makes for a heavy gate. To avoid sagging and to ease the gate's swing, you'll need to include a wheel on the latch side of the gate. Over a solid surface such as concrete or asphalt, the wheel will help you open and close the gate easily.

Shaping the top of the arch is a simple matter: just enlarge the pattern provided on page 127 and trace it onto the siding. Then cut the shape using a jigsaw.

This piece of wrought iron came from a banister we found at a salvage yard. We used a reciprocating saw with a metal-cutting blade to cut it to a usable size.

**This decorative piece of wrought iron** salvaged from a banister adds intrigue and a functional peephole to this classic gate shape.

# Tools, Materials & Cutting List

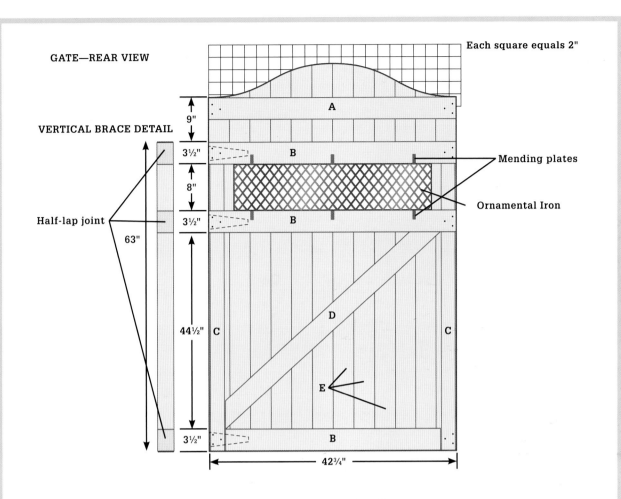

GATE—REAR VIEW

Each square equals 2"

VERTICAL BRACE DETAIL

Half-lap joint

9"

3½"

8"

3½"

63"

44½"

3½"

A

B

B

C

D

E

B

C

42¾"

Mending plates

Ornamental Iron

Tape measure
Circular saw w/ wood & metal cutting blade
Chisel
Drill/driver and bits
Jigsaw
Level
Framing square
Spring clamps
Caulk gun
Salvaged piece of ornamental metal
Pressure-treated KDAT (Kiln Dried After
    Treatment) cedar or redwood lumber:
    2 × 4s, 10 ft. (3); 1 × 4s, 8 ft. (13)
Posterboard or cardboard
Construction adhesive
Corrosion-resistant deck screws (1¼", 2", 4")
1½" corrosion-resistant mending plates (8)
2½" corrosion-resistant bolts and nuts (8)
Gate wheel
Hinge & latch hardware

Gate handle
Exterior wood glue
Mending plates
Ratchet wrench
Screwdriver
Wood shims
Hammer
Eye and ear protection
Work gloves

| Key | Part | Type | Size | Number |
|-----|------|------|------|--------|
| A | Siding brace | 1 × 4 | 42¾" | 1 |
| B | Horizontal braces | 2 × 4 | 42¾" | 3 |
| C | Vertical braces | 2 × 4 | 63" | 2 |
| D | Diagonal brace | 2 × 4 | 6 ft. | 1 |
| E | Siding | 1 × 4 | 8 ft. | 12 |

# How to Build an Arched Gate

**Measure the opening between your gate posts** and determine the finished size of the gate (check the packaging of the hinge and latch hardware for recommended clearances). The finished gate in this project is 42¾" wide, but you can modify any dimensions as needed to suit your project. Cut all pieces for the gate according to the cutting list on page 127.

**Cut the half-lap joints** using a circular saw and chisel. These appear on both ends of the horizontal braces and at the three locations on the vertical braces, as shown on page 127. First make the shoulder cuts with a circular saw set for a ¾"-deep cut. Then make a series of closely spaced cuts to the end of the board (or the other shoulder cut). Remove the waste and smooth the bottoms of the notches with a chisel. Test-fit the cuts and make any necessary adjustments.

**Assemble the brace frame.** Fit the pieces together back-side-up, and measure diagonally from corner to corner to check for square (when the diagonals match, the frame is square). Secure the half-lap joints with exterior wood glue and 1¼" deck screws. Position the diagonal brace as shown (with its bottom end at the hinge-side of the frame) and mark the ends for cutting. Make the cuts, and fasten the diagonal brace to the vertical braces with 4" deck screws driven through angled pilot holes. Finish the wood, if necessary.

**Add the siding boards and siding brace.** Screw the two outer siding boards to the brace frame using pairs of 2" deck screws. The boards should be flush with the outside edges and bottom of the brace frame. Install the remaining boards in between, spacing them evenly as desired. On the back side, fasten the siding brace across the siding with glue (or construction adhesive) and 1¼" screws so its top edge is 9" above the top of the top horizontal brace.

**Cut the display opening.** Set your display piece onto the front of the gate so it is centered within the framed opening (you can drill small holes to locate the braces), and trace its outline onto the siding. Cut inside the outline with a circular saw set just deeper than ¾", and then finish the corner cuts with a jigsaw or chisel.

**Shape the gate top.** First create a cardboard template of the arch shape: either enlarge the image on page 127 on a photocopier, or draw the shape using the grid as a guide. Trace the template shape onto the siding (front side) so the ends of the arch are even with the siding brace. Make the cut with a jigsaw and a down-cutting blade, cutting from the front side to prevent splintering. Finish all wood, as necessary.

**Install the display piece,** using small mending plates with two or more holes. Position the plates evenly around the edges of the piece, and mark the piece for bolt holes. Drill the bolt holes (for iron, start with a small bit, then drill with successively larger bits; use a slow drill speed throughout). Set the piece and plates in place on the gate and mark for bolt holes. Drill the holes and anchor the piece with bolts and nuts.

**Hang the gate.** Mount the hinges onto the gate. Ideally, the hinge leaves should mount to the three 2 × 4 horizontal braces of the gate frame. Also install the gate wheel. Set the gate into position, and shim along the bottom and latch side to create the desired gaps. Make sure the gate is plumb, and fasten the hinges to the hinge post. Install the latch hardware, as desired.

# Trellis Gate

**A trellis gate** is a perfect home for climbing flowers and vines, creating a lovely portal to your home or garden.

This trellis gate combination is a grand welcome to any yard. But don't let its ornate appearance fool you—the simple components create an impression far beyond the skills and materials involved in its construction.

This gate is best suited to a location where it will receive plenty of sunlight to ensure an abundant canopy of foliage. Choose perennials rather than annuals, since they will produce more luxurious growth over time. Heirloom roses are a good choice, providing a charming complement to the gate's old-fashioned look and air of elegance.

Larger, traditional styles of hardware that showcase well against the painted wood will also enhance the gate's impressive presentation. The hardware and the millwork we used are available at most building centers, but you might want to check architectural salvage shops. They may have unique pieces that add special character to the gate.

As with most of our projects, you can alter the dimensions of this project to fit an existing opening. Just recalculate the materials and cutting lists, and make sure you have enough lumber to accommodate the changes.

## Tools, Materials & Cutting List ▸

Tape measure
Circular saw
Paintbrush
Drill
Carpenter's level
Framing square
Jigsaw
Chisel
Hand maul
Hammer
Wood shims
Pressure-treated cedar
  or redwood lumber:
  2 × 2s, 8 ft. (8);
  2 × 4s, 8 ft. (9);
  1 × 4s, 8 ft. (4);
  1 × 6, 8 ft. (1);
  1 × 2, 4 ft. (1)
Galvanized
  deck screws (1¼",
  2½", 1½", 2", 3")

Galvanized finish nails
24" pressure-treated
  stakes (4)
Galvanized lag screws
  (2", 3")
Cardboard or
  posterboard
Victorian millwork
  brackets (4)
Hinge hardware
Gate handle
Flexible PVC pipe
Clamps
Ratchet wrench
Eye and ear protection
Work gloves
Primer
Framing square
Paint

| Part | Type | Size | Qty. |
|---|---|---|---|
| **Frames** | | | |
| Horizontal braces | 2 × 2 | 12" | 2 |
| | 2 × 2 | 15¾" | 8 |
| | 2 × 2 | 33" | 6 |
| Vertical braces | 2 × 2 | 17" | 4 |
| | 2 × 2 | 54½" | 2 |
| | 2 × 4 | 87½" | 4 |
| Stop | 1 × 2 | 46½" | 1 |
| **Top** | | | |
| Tie beams | 2 × 4 | 72¾" | 2 |
| Rafters | 2 × 2 | 33" | 4 |
| **Gate** | | | |
| Hor. braces | 2 × 4 | 40½" | 2 |
| Vert. braces | 2 × 4 | 32¾" | 2 |
| Diag. brace | 2 × 4 | 49½" | 1 |
| Siding | 1 × 4 | 45¼" | 7 |
| | 1 × 6 | 45¼" | 2 |

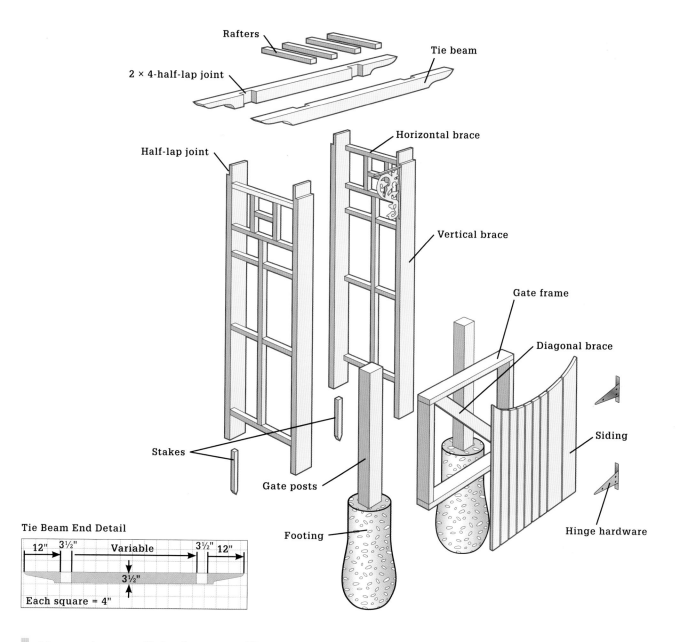

Rafters

2 × 4-half-lap joint

Tie beam

Half-lap joint

Horizontal brace

Vertical brace

Gate frame

Diagonal brace

Siding

Stakes

Gate posts

Footing

Hinge hardware

**Tie Beam End Detail**

12"   3½"   Variable   3½"   12"

3½"

3½"

Each square = 4"

## How to Build the Trellis Gate

**1**

**Build the trellis frames.** Cut the horizontal and vertical braces for each trellis frame, following the cutting list on page 130. Cut notches for the half-lap joints on the top end of each 2 × 4 vertical brace (see page 128, step 2). The notches are 3½" long × ¾" deep. Paint the parts with primer and then lay them on a flat surface. Assemble the frames with 2½" deck screws. Use a framing square to make sure the parts are square before fastening.

(continued)

**Set gate posts in concrete** (see pages 26 to 29) and then anchor the frames to the gate posts. Position each frame against the back face of a post so the inside edge of the frame is flush. Check to make sure the frame is plumb. Fasten the frame brace to the post with 3" lag screws driven through pilot holes.

**Secure the free ends of the frames.** Measure diagonally between opposing corners of the frames, and move the free ends of the frames as needed until the diagonals are equal. Drive 24"-long 1 × 2 or 2 × 2 stakes (pressure-treated) behind the rear vertical brace of each frame, and fasten the stakes to the braces with 2" lag screws driven through pilot holes (or use 2½" screws for 2 × 2 stakes).

**Cut and install the tie beams.** Measure across the frame braces to find the overall length of the tie beams. Make a cardboard template of the Tie Beam End Detail using the drawing on page 131 as a guide. Cut the tie beams to length. Hold them in place against the frame braces and mark the half-lap joints. Cut the half-lap notches on the beams. Shape the ends of the beams with a jigsaw, using the template to draw the outline. Fasten the beams to the braces with 1¼" deck screws.

**Install the rafters.** Make layout marks for the four rafters on the inside faces of the tie beams, spacing them evenly between the side frames. Cut the rafters to fit snugly between the beams. Fasten the rafters to the beams with 3" deck screws driven through pilot holes.

**Add the trim.** Position a decorative bracket at each corner of the tie beam and vertical brace, and fasten it in place with appropriately sized finish nails driven through pilot holes. Add brackets at all four corners.

**Construct the gate frame.** Cut the horizontal and vertical braces for the gate, following the cutting list on page 130 (or modify the dimensions to fit your project). Assemble the parts with 2½" deck screws. Check the frame for square. Set the diagonal brace under the frame and mark the end cuts, as shown here. Cut the brace and fasten it to the frame with screws.

**Install the gate siding.** Cut the siding boards to length. Clamp a 1 × 4 spacer to the bottom of the gate frame. Install the siding flush with the edge of the 1 × 4, starting with two 1 × 6s on the hinge-side of the gate. Gap the siding with ⅝" spacers, and fasten the siding to the gate frame with pairs of 2" deck screws. Hang the gate between the gate posts.

**Cut the gate siding top profile.** Drive a nail in the center of the gate, a little bit above the top horizontal brace. Bend a length of flexible PVC pipe under the nail and clamp the ends to the top outside corners of the siding boards. Mark along the pipe to draw the curve. Cut the boards with a jigsaw. Touch up primer and paint the gate trellis.

# Garden Walls

Like fences, walls for the home landscape can be crafted from a broad range of materials—from ancient building blocks, such as natural stone and clay brick, to clever modern products, like molded concrete units designed for easy, mortarless installation. Your project might also call for poured concrete, either in a structural footing for the wall or to create the wall itself. Whichever material you choose, you'll find that each type of wall offers not only a distinct look and feel, but also an opportunity to learn a special skill, making the building process almost as rewarding as the finished product.

As a general rule of thumb, masonry walls built by amateurs should be kept at about three feet or shorter. The same is true for all retaining walls. The reason is that the taller a wall is, the more support it needs—either from the base, special stacking techniques, internal reinforcement, or a combination of elements. Building in this support requires expert knowledge of the specific material, and often some engineering calculations. At three feet and under, walls have a relatively low center of gravity and are less subject to wind and other forces. If you need a tall barrier, you should consider hiring a professional to build the wall, or you can opt to build a tall fence instead.

### In this chapter:

- Patio Wall
- Outdoor Kitchen Walls & Countertop
- Dry Stone Wall
- Mortared Stone Wall
- Brick Garden Wall
- Mortarless Block Wall
- Poured Concrete Wall
- Interlocking Block Retaining Wall
- Timber Retaining Wall
- Stone Retaining Wall
- Poured Concrete Retaining Wall

# Patio Wall

Perhaps due to the huge popularity of interlocking concrete wall block, which made building retaining walls a great do-it-yourself project, you can now find concrete landscape blocks made for a range of applications, including patio walls, freestanding columns, raised planters, and even outdoor kitchens (see page 142). The blocks shown in this project require no mortar and are stacked up just like retaining wall units. Yet unlike retaining wall blocks, these "freestanding" units have at least two faceted faces, so the wall looks good on both sides. And they have flat bottoms, allowing them to be stacked straight up without a batter (the backward lean required for most retaining walls).

Solid concrete blocks for freestanding walls come in a range of styles and colors. Products that come in multiple sizes produce walls with a highly textured look that mimics natural stone, while walls made with uniform blocks have an appearance closer to weathered brick. Many block products can be used for both curved and straight walls, and most are compatible with cap units that give the wall an architecturally appropriate finish, as well as a great surface for sitting.

The wall in this project forms a uniform curve to follow the shape of a circular patio. It's built over a base of compacted gravel, but you could build the same wall right on top of a concrete patio slab. Keep in mind that freestanding walls like this are typically subject to height limits, which might range from 20 to 36 inches or higher. Walls with straight sections or gentle curves may need a supporting feature, such as a column or 90-degree turn or jog, to stabilize the wall.

## Tools & Materials ▸

| | |
|---|---|
| Mason's string | Wood stakes |
| Landscape marking paint | Straight board |
| | Compactable gravel |
| Excavation tools | Concrete wall block |
| Line level | and cap units |
| Plate compactor or hand tamp | Concrete adhesive |
| | Tape measure |
| Rake | Eye and ear protection |
| 4-ft. level | Circular saw with masonry |
| Caulk gun | blade (optional) |
| Brickset or pitching chisel | Heavy rope or garden hose |
| | Wheelbarrow |
| Hand maul | Work gloves |
| Stone chisel | |

**Landscape block** for freestanding walls is versatile and an easy material with which to build. You can use it to create low walls of almost any shape, plus columns, steps, and other features. Quality block manufacturers offer a variety of styles and textures, along with compatible specialty and accessory pieces for a well-integrated look.

# Laying Out Freestanding Block Walls

**Draw the rough outline of the wall** onto the ground with a can of marking paint. First measure the wall blocks and/or align a few blocks in place as guides. To mark end columns, first measure the blocks and then use the marking paint to outline the footprint of the column (inset).

**Freeform curving walls:** Use heavy rope or a garden hose to lay out the wall's shape. Follow the rope with marking paint to transfer the outline to the ground. To mark the other side of the wall and the edges of the excavation, reposition the rope or hose the appropriate distance away from the first mark and trace with paint.

**Straight walls:** Mark the outlines of the wall and/or excavation with stakes and mason's string. Position one string, then measure from it to position any remaining strings as needed. *Tip: Leave the stakes marking one of the wall faces in the ground; you'll use them later to align the wall block.*

# How to Build a Freestanding Patio Wall

**Remove the sod and other plantings** inside the excavation area. For a gravel base, the excavation should extend 6" beyond the wall on all sides. If you are building adjacent to a sandset patio with pavers, take care not to disturb the rigid paver edging. Alternatively, fully excavate the ground around patio to compensate for wall addition and install new edging around perimeter. Follow your manufacturer's instructions.

**Set up level lines** to guide the excavation using stakes and mason's string. For curved walls, you may need more than one string. Level the string with a line level (make sure multiple strings are level with one another). Measure from the string to ground level (grade), and then add 12" (or as directed by the block manufacturer)—this is the total depth required for the excavation.

**Use a story pole** to measure the depth as you complete the excavation. To make a story pole, mark the finished depth of the excavation onto a straight board, and use it to measure against the string; this is easier than pulling out your tape measure for each measurement.

**Tamp the soil in the trench** with a rented plate compactor or a hand tamp. The bottom of the trench should be flat and level, with the soil thoroughly compacted. Take care not to disturb or damage adjacent structures.

**Spread compactable gravel** over the trench in an even 2- to 3"-thick layer. Tamp the gravel thoroughly. Add the remaining gravel and tamp to create a 6"-thick layer after compaction.

**Check the gravel base with a level** (or a level taped to a straight board) to make sure the surface is uniform and perfectly level. Add gravel to any low spots and tamp again.

**Set the first course.** If you're using more than one thickness of block, select only the thicker units for the first course. Lay out the blocks in the desired pattern along the layout line, butting the ends together for complete contact. If necessary, cut blocks to create the desired curve (see step 10). Place a 4-ft. level across the blocks to make sure they are level and flat across the tops.

**Set the second course.** Begin the course at the more visible end of the wall. Set the blocks in the desired pattern, making sure to overlap the block joints in the first course to create a bond pattern. Alternate different sizes of block regularly, and check the entire course with a level. If necessary, cut a block for the end of the wall.

(continued)

**9**

**End each course** with a piece no narrower than 6". If necessary, position a full unit at the end of the wall, then measure back and cut the second-to-last unit to fit the space. Glue small end pieces in place with concrete adhesive.

**10**

**Cut blocks** using a brickset or pitching chisel and a maul. First score along the cutting line all the way around the block, and then chisel at the line until the block splits. You can also cut a deep score line (on thick block) or cut completely through (on thin block) using a circular saw with a masonry blade.

**Cut Facets ▸**

**Round over the cut edges** of blocks to match the original texture. Using a stone chisel and mason's hammer or maul, carefully chip along the edges to achieve the desired look.

**11**

**Complete the remaining courses,** following the desired pattern. Be sure to maintain a bond with the course below by overlapping the joints in the lower course. For the top two courses, glue each block in place with concrete adhesive.

**12**

**Install the cap blocks.** Trapezoidal cap block may fit your wall's curve well enough without cuts (for gentle curves, try alternating the cap positions). If cuts are necessary, dry-fit the pieces along the wall, and plan to cut every other block on both side edges for an even fit. Set all caps with concrete adhesive. Backfill along the wall to bury most or all of the first course.

# How to Add Decorative Columns to a Wall

**Set the first course** of each column after completing the first wall course (middle-of-wall columns are set along with each wall course). Use four full blocks for the first course, butting the column blocks against the end wall block. Check the column blocks for level. *Note: Prepare the ground as seen on pages 137 to 138.*

**Glue the second course** and all subsequent courses in place with concrete adhesive or according to the manufacturer's specifications.

**Cap the column with special cap units,** or create your own caps with squares of flagstone. Glue cap pieces in place with concrete adhesive or mortar in between them, following the manufacturer's instructions. *Tip: The hollow space in the column's center is ideal for running wiring for adding a light fixture on top of a cap.*

# Outdoor Kitchen Walls & Countertop

Loaded with convenient work surfaces and a dedicated grill space, the outdoor kitchen has changed backyard grilling forever. This roomy kitchen can be the perfect addition to any patio or garden retreat. It's made entirely of concrete blocks and not only looks great, it's also incredibly easy to build.

The design of this kitchen comes from a manufacturer (see Resources, page 204) that supplies all of the necessary masonry materials on two pallets. As shown, the project's footprint is about 98 × 109 inches and includes a 58-inch-wide space for setting in a grill. Square columns can provide work surfaces on either side of the grill, so you'll want to keep them conveniently close, but if you need a little more or a little less room for your grill, you can simply adjust the number of blocks that go into the front wall section enclosing the grill alcove.

Opposite the grill station is a 32-inch-tall countertop capped with large square pavers, or patio stones, for a finished look. This countertop has a lower surface for food prep and a higher surface for serving or dining. A low side wall connects the countertop with the grill area and adds just the right amount of enclosure to complete the kitchen space.

## Tools & Materials ▸

Masonry outdoor kitchen kit (concrete wall block, concrete patio stones)
Chalk line
Framing square
Straight board
Level
Caulk gun
Exterior-grade concrete adhesive
Tape measure
Eye and ear protection
Work gloves

**This all-masonry outdoor kitchen** comes ready to assemble on any solid patio surface, or you can build it over a prepared gravel base anywhere in your landscape (check with the manufacturer for base requirements). For a custom design, similar materials are available to purchase separately and the installation would be more or less the same as shown here. Discuss the project with the manufacturer for specifics. If you decide to build just a part of this kitchen (the bar, for example), review the setup and site prep steps at the beginning of this project.

# How to Build the Outdoor Kitchen

**Dry-lay the project** on the installation surface. This overview of the first course of blocks shows how the kitchen is constructed with five columns and two wall sections. Laying out the first course carefully and making sure the wall sections are square ensures the rest of the project will go smoothly.

Side wall

Long wall

Short wall

**Create squared reference lines** for the kitchen walls after you remove the dry-laid blocks. Snap a chalk line representing the outside face of the front wall. Mark the point where the side wall will meet the front wall. Place a framing square at the mark and trace a perpendicular line along the leg of the square. Snap a chalk line along the pencil line to represent the side wall, or use the edge of a patio as this boundary (as shown). To confirm that the lines are square, mark the front-wall line 36" from the corner and the side-wall line 48" from the corner. The distance between the marks should be 60". If not, re-snap one of the chalk lines until the measurements work out.

(continued)

**Begin laying the first course of block.** Starting in the 90° corner of the chalk lines, set four blocks at right angles to begin the corner column. Make sure all blocks are placed together tightly. Set the long wall with blocks laid end to end, followed by another column.

**Finish laying the first course,** including two more columns, starting at the side wall. Use a straight board as a guide to make sure the columns form a straight line. To check for square, measure between the long wall and the short wall at both ends; the measurements should be equal. Adjust the short-wall columns as needed.

**Set the second course.** Add the second course of blocks to each of the columns, rotating the pattern 90° to the first course. Set the blocks for the long and side walls, leaving about a 2" gap in between the corner column and the first block. Set the remaining wall blocks with the same gap so the blocks overlap the joints in the first course.

**Set the third course.** Lay the third-course blocks using the same pattern as in the first course. For appearance and stability, make sure the faces of the blocks are flush with one another and that the walls and columns are plumb. Use a level to align the blocks and check for plumb.

**Install the remaining courses.** The higher courses of wall block are glued in place. Set the courses in alternating patterns, as before, gluing each block in place with concrete adhesive.

**Build the short wall overhang.** Starting at one end of the short wall, glue wall blocks along the tops of the columns with concrete adhesive. Position blocks perpendicular to the length of the short wall, overhanging the columns by 3".

**Complete the short wall top.** Create the counter surface for the short wall by gluing patio stones to the tops of the columns and overhanging blocks. Position the stones for the lower surface against the ends of the overhanging blocks. Position the upper-surface stones so they extend beyond the overhanging blocks slightly on the outside ends and a little more so on the inside ends.

**Cap the corner columns.** Finish the two corner columns with wall blocks running parallel to the side wall. Glue the cap pieces in place on the colomns using concrete adhesive. Make sure the blocks are fitted tightly together.

# Dry Stone Wall

You can construct a low stone wall without mortar, using a centuries-old method known as "dry laying." With this technique, the wall is actually formed by two separate stacks that lean together slightly. Each stone overlaps a joint in the previous course. This technique avoids long vertical joints, resulting in a wall that is attractive and strong. The position and weight of the two stacks support each other, forming a single, sturdy wall.

While dry walls are simple to construct, they do require a fair amount of patience. The stones must be carefully selected and sorted by size and shape. They must also be correctly positioned in the wall so that weight is distributed evenly. Long, flat stones work best. A quarry or aggregate supply center will have a variety of sizes, shapes, and colors to choose from. For this project you'll need to purchase a number of stones in these four sizes:

- Shaping: half the width of the wall
- Tie: the same width as the wall
- Filler: small shims that fit into cracks
- Cap: large, flat stones, wider than the wall

Because the wall relies on itself for support, a concrete footing is unnecessary, but the wall must be at least half as wide as it is tall. This means some stones may need to be shaped or split to maintain the spacing and structure of the wall. See pages 30 to 33 for tips on working with stone.

## Tools & Materials ▶

| | |
|---|---|
| Circular saw with | Type M mortar |
| masonry blade | Rough-textured rag |
| Hand sledge | Compactable gravel |
| Mason's chisel | Excavation tools |
| 4-ft. level | Mason's string |
| Mason's trowel | Wood stakes |
| Stones of various | Tape measure |
| shapes and sizes | Eye and ear protection |
| Cap stones | Work gloves |

**A dry stone wall** is one of the oldest and strongest styles of garden wall out there. The wall's two stacks of stones rely on one another for support.

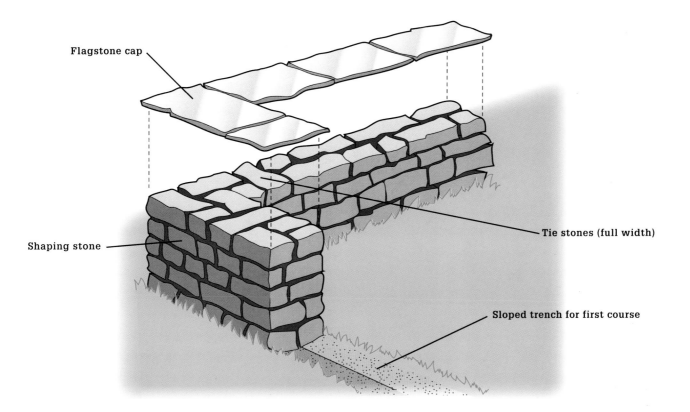

Flagstone cap

Shaping stone

Tie stones (full width)

Sloped trench for first course

## How to Build a Dry Stone Wall

**1**

**Lay out the wall site with stakes** and mason's string. Dig a 24"-wide trench that is 6" deep at the edges and 8" deep in the center, creating a slight V shape by evenly sloping the sides toward the center. Compact any loose soil. Add a 2"-layer of gravel, but do not compact it.

**2**

Shaping stones (½ wall width)

**Lay two rows of shaping stones** along the bottom of the trench. Position them flush with the edges of the trench and sloping toward the center, staggering joints. Use stones similar in height. If stones have uneven surfaces, position them with the uneven sides facing down.

**3**

**Form a corner** by laying the last stone of the outer row so it covers the end of the stone in the outer row of the adjacent wall course. Lay the inner row in the same manner.

(continued)

**Lay the second course** and fill any significant gaps between the shaping stones with rubble and filler stones.

Filler stones

Tie stone

**Lay the stones for the second course corner** so they cover the joints of the first course corner. Form corners using the same steps as for forming the first course corner. Use stones that have long, square sides. Place tie stones across the width of each wall just before the corner. Build the wall ends in this same way. Use stones of varying lengths so that each joint is covered by the stone above it. Wedge filler stones into any large gaps.

**Lay the third course.** Work from the corner to the end of the wall. If necessary, shape or split the final stones of the course to size with a masonry saw or hand sledge and chisel. Place tie stones approximately every 36". Lay shaping stones between the tie stones. Make sure to stagger the joints; stones of varying lengths will help offset them. Continue to place filler stones into any cracks on the surface or sides of the wall. Continue laying courses, maintaining a consistent height along the wall and adding tie stones to every third course. Check for level as you go.

**When the wall is about 36" high,** check for level. Trowel mortar onto the center of the wall, in at least 6" from the edges. Center the capstones and set them as close together as possible. Carefully fill the cracks between the capstones with mortar. Let any excess mortar dry until crumbly, then brush it off. After two or three days, scrub off any residue using water and a rough-textured rag.

If slope is an issue along your wall site, you can easily build a stepped wall to accommodate it. The key is to keep the stones level so they won't shift or slide with the grade, and to keep the first course below ground level. This means digging a stepped trench.

Lay out the wall site with stakes and mason's string. Dig a trench 4 to 6" deep along the entire site, including the slope. Mark the slope with stakes at the bottom where it starts, and at the top where it ends.

Begin the first course along the straight-line section of the trench, leading up to the start of the slope. At the reference stake, dig into the slope so a pair of shaping stones will sit level with the rest of the wall.

To create the first step, excavate a new trench into the slope, so that the bottom is level with the top of the previous course. Dig into the slope the length of one-and-a-half stones. This will allow one pair of stones to be completely below the ground level, and one pair to span the joint where the new trench and the stones in the course below meet.

Continue creating steps, to the top of the slope. Make sure each step of the trench section remains level with the course beneath. Then fill the courses, laying stones in the same manner as for a straight-line wall. Build to a maximum height of 36", and finish by stepping the top to match the grade change, or create a level top with the wall running in to the slope.

If you'd like a curved wall or wall segment, lay out each curve, as demonstrated on page 137. Then dig the trench as for a straight wall, sloping the sides into a slight V toward the center. Lay the stones as for a straight wall, but use shorter stones; long, horizontal stones do not work as well for a tight curve. Lay the stones so they are tight together, offsetting the joints along the entire stretch. Be careful to keep the stone faces vertical to sustain the curve all the way up the height of the wall.

**To build a curved wall,** lay out the curve using a string staked to a center point as a compass. Then, dig the trench and set stones using the same techniques as for a straight wall.

**If the wall goes up- or downhill,** step the trench, the courses, and the top of the wall to keep the stones level.

# Mortared Stone Wall

## Tools & Materials ▶

Tape measure
Pencil
Chalk line
Small whisk broom
Tools for
    mixing mortar
Trowel
Jointing tool
Line level
Sponge
Garden hose
Concrete materials
    for footing

Ashlar stone
Type N or
    Type S mortar
Stakes and
    mason's line
Scrap wood
Muriatic acid
4-ft. level
Protective clothing
Bucket
Eye and
    ear protection
Work gloves

The mortared stone wall is a classic that brings structure and appeal to any yard or garden. Square-hewn ashlar is the easiest to build with, though field stone and rubble also work and make attractive walls.

Because the mortar turns the wall into a structure that can crack and heave with the freeze-thaw cycle, a concrete footing is required for a mortared stone wall. To maintain strength in the wall, use the heaviest, thickest stones for the base of the wall and thinner, flatter stones for the cap.

As you plan the wall layout, install tie stones—stones that span the width of the wall—about every three feet, staggered through the courses both vertically and horizontally throughout the wall. Use the squarest, flattest stones to build the "leads," or ends of the wall first, then fill the middle courses. Plan for joints around one inch thick, and make sure joints in successive courses do not line up.

Laying a mortared stone wall is labor-intensive, but satisfying work. Make sure to work safely and enlist friends to help with the heavy lifting. See pages 30 to 33 for tips on working with stone.

**A mortared stone wall** made from ashlar adds structure and classic appeal to your home landscape. Plan carefully and enlist help to ease the building process.

# How to Build a Mortared Stone Wall

**Pour a footing for the wall** and allow it to cure for one week (pages 42 to 45). Measure and mark the wall location so it is centered on the footing. Snap chalk lines along the length of the footing, for both the front and the back faces of the wall. Lay out corners using the 3-4-5 right angle method as described on page 24.

**Dry-lay the entire first course.** Starting with a tie stone at each end, arrange stones in two rows along the chalk lines with joints about 1" thick. Use smaller stones to fill the center of the wall. Use larger, heavier stones in the base and lower courses. Place additional tie stones approximately every 3 feet. Trim stones as needed.

**Mix a stiff batch of Type N or Type S mortar,** following the manufacturer's directions (see pages 38 to 39). Starting at an end or corner, set aside some of the stone and brush off the foundation. Spread an even, 2"-thick layer of mortar onto the foundation, about ½" from the chalk lines—the mortar will squeeze out a little.

**Firmly press the first tie stone into the mortar,** so it is aligned with the chalk lines and relatively level. Tap the top of the stone with the handle of the trowel to set it. Continue to lay stones along each chalk line, working to the opposite end of the wall.

(continued)

**After installing** the entire first course, fill voids along the center of the wall that are larger than 2" with smaller rubble. Fill the remaining spaces and joints with mortar using the trowel.

**As you work,** rake the joints using a scrap of wood to a depth of ½"; raking joints highlights the stones rather than the mortared joints. After raking, use a whisk broom to even the mortar in the joints.

**Variation:** You can also tool joints for a cleaner, tighter mortared joint. Tool joints when your thumb can leave an imprint in the mortar without removing any it.

**Drive stakes at the each end of the wall** and align a mason's line with the face of the wall. Use a line level to level the string at the height of the next course. Build up each end of the wall, called the "leads," making sure to stagger the joints between courses. Check the leads with a 4-ft. level on each wall face to verify plumb.

**If heavy stones push out too much mortar,** use wood wedges cut from scrap to hold the stone in place. Once the mortar sets up, remove the wedges and fill the voids with fresh mortar.

## Clean Up Spills ▸

**Have a bucket of water and a sponge handy** in case mortar oozes or spills onto the face of the stone. Wipe mortar away immediately before it can harden.

**Fill the middle courses between the leads** by first dry laying stones for placement and then mortaring them in place. Install tie stones about every 3 feet, both vertically and horizontally, staggering their position in each course. Make sure joints in successive courses do not fall in alignment.

**Install cap stones** by pressing flat stones that span the width of the wall into a mortar bed. Do not rake the joints, but clean off excess mortar with the trowel and clean excess mortar from the surface of the stones using a damp sponge.

**Allow the wall to cure for one week,** then clean it using a solution of 1 part muriatic acid and 10 parts water. Wet the wall using a garden hose, apply the acid solution, then immediately rinse with plenty of clean, clear water. Always wear goggles, long sleeves and pants, and heavy rubber gloves when using acids.

# Brick Garden Wall

Traditional brick construction is a timeless symbol of craftsmanship and architectural permanence. With its solid yet warmly textured appearance, a low brick wall makes a great backdrop for a garden or a partition or accent wall for a patio.

The wall in this project is freestanding and built with "double-wythe" construction (with two layers of brick tied together), which is necessary for support. Other brick projects, such as a square or rectangular planter, can be built with single-wythe construction, as long as the shape of the structure provides its own support.

All brick structures must be built upon a solid concrete footing (at least twice as wide as the finished wall) or a suitable concrete slab. Consult your city's building department to learn about permits and structural requirements for your footing and wall. Often, specifications are less stringent for walls that are under a certain height, usually three feet.

See pages 40 to 45 for steps on building a concrete footing, and pages 34 to 39 for tips on working with brick and mortar.

**Brick and mortar partition walls** help define space in your landscape and add timeless sophistication to your home's appearance.

## Tools & Materials ▸

| | | | |
|---|---|---|---|
| Work gloves | Line blocks | Brick | ⅜"-dia. dowel |
| Trowel | Mason's string | Wall ties | Tools for cutting brick |
| Chalk line | Jointing tool | Rebar (optional) | Eye and ear protection |
| Level | Mortar | | |

## How to Build a Brick Garden Wall

**Dry-lay the first course** by setting down two parallel rows of brick, spaced ¾ to 1" apart. Use a chalk line to outline the location of the wall on the slab. Draw pencil lines on the slab to mark the ends of the bricks. Test-fit the spacing with a ⅜"-diameter dowel, then mark the locations of the joint gaps to use as a reference after the spacers are removed.

**Dampen the concrete slab or footing with water,** and dampen the bricks or blocks if necessary. Mix mortar and throw a layer of mortar onto the footing for the first two bricks of one wythe at one end of the layout. Butter the inside end of the first brick, then press the brick into the mortar, creating a ⅜" mortar bed. Cut away excess mortar.

**Plumb the face of the end brick** using a level. Tap lightly with the handle of the trowel to correct the brick if it is not plumb. Level the brick end to end. Butter the end of a second brick, then set it into the mortar bed, pushing the dry end toward the first brick to create a joint of ⅜".

**Butter and place a third brick,** using the chalk lines as a general reference, then check for level and plumb. Adjust any bricks that are not aligned by tapping lightly with the trowel handle.

**Lay the first three bricks** for the other wythe parallel to the first wythe. Level the wythes, and make sure the end bricks and mortar joints align. Fill the gaps between the wythes at each end with mortar.

**Cut a half brick,** then throw and furrow a mortar bed for a half brick on top of the first course. Butter the end of the half brick, then set it in the mortar bed, creating a ⅜" joint. Cut away excess mortar. Check bricks for plumb and level.

Header bracks

**Add more bricks and half bricks** to both wythes at the end until you lay the first bricks in the fourth course. Align bricks with the reference lines. *Note: To build corners, lay a header brick at the end of two parallel wythes. Position the header brick in each subsequent course perpendicular to the header brick in the previous course (inset).*

(continued)

**Check the spacing of the end bricks** with a straightedge. Properly spaced bricks will form a straight line when you place the straightedge over the stepped end bricks. If bricks are not in alignment, do not move those bricks already set. Try to compensate for the problem gradually as you fill in the middle (field) bricks (step 11) by slightly reducing or increasing the spacing between the joints.

**Every 30 minutes,** stop laying bricks and smooth out all the untooled mortar joints with a jointing tool. Do the horizontal joints first, then the vertical joints. Cut away any excess mortar pressed from the joints using a trowel. When the mortar has set, but is not too hard, brush any excess mortar from the brick faces.

Line block

**Build the opposite end of the wall** with the same methods as the first, using the chalk lines as a reference. Stretch a mason's string between the two ends to establish a flush, level line between ends—use line blocks to secure the string. Tighten the string until it is taut. Begin to fill in the field bricks (the bricks between ends) on the first course, using the mason's string as a guide.

**Lay the remaining field bricks.** The last brick, called the closure brick, should be buttered at both ends. Center the closure brick between the two adjoining bricks, then set in place with the trowel handle. Fill in the first three courses of each wythe, moving the mason's string up one course after completing each course.

Metal wall tie

**In the fourth course,** set metal wall ties into the mortar bed of one wythe and on top of the brick adjacent to it. Space the ties 2 to 3 ft. apart, every three courses. If desired (or required by code), strengthen the wall by setting metal rebar into the cavities between the wythes and filling with thin mortar per code specifications..

**Lay the remaining courses,** installing metal ties every third course. Check with mason's string frequently for alignment, and use a level to make sure the wall is plumb and level.

**Lay a furrowed mortar bed on the top course,** and place a wall cap on top of the wall to cover empty spaces and provide a finished appearance. Remove any excess mortar. Make sure the cap blocks are aligned and level. Fill the joints between cap blocks with mortar.

# Mortarless Block Wall

Far from an ordinary concrete block wall, this tile-topped mortarless block wall offers the advantages of concrete block—affordability and durability—as well as a dramatic touch of style. Color is the magic ingredient. Tint added to the surface-bonding cement produces a buttery-yellow that contrasts beautifully with the cobalt blue tile. Of course, you can use any color combination that matches or complements your setting.

Mortarless block walls are simple to build. You set the first course in mortar on a concrete footing and stack the subsequent courses in a running bond pattern, without mortar between the blocks. The wall gets its strength from a coating of surface-bonding cement that's applied to the exposed surface. Tests have shown that the bond created in this type of construction is just as strong as traditional block-and-mortar walls.

In this project, the wall is 24 inches tall and uses three courses of standard 8 × 8 × 16-inch concrete blocks and 8 × 12-inch ceramic tiles for the top cap, with bullnose tiles to finish the edges. When selecting tile, be sure to get durable exterior ceramic tile, thinset exterior tile mortar, and exterior tile grout.

The footing below the wall must be twice as wide as the wall and extend at least 12 inches beyond each end (see pages 40 to 45 for instructions on building a concrete footing). Consult with your local building department to learn about footing size and depth, maximum wall height, and other structural requirements for this type of wall. You may also be allowed to build the wall on top of an approved concrete slab.

## Tools & Materials ▶

| | |
|---|---|
| Line level | Circular saw with |
| Mortar box | masonry-cutting blade |
| 4-ft. level | Surface-bonding cement |
| Chalk line | 8 × 12" ceramic tile |
| Masonry chisel | Matching bullnose tile |
| Line blocks | Sand-mix exterior grout |
| Mason's trowel | Grout sealer |
| Notched trowel | Tape measure |
| Tile cutter | Spray bottle |
| Rubber grout float | Mason's string |
| Sponge | Coloring agent |
| Concrete blocks | Finishing trowel |
| Type N mortar | Mortar hawk |
| Wire mesh | Masking tape |
| Thinset mortar | Work gloves |
| Tile spacers | Eye and ear protection |

**The mortarless block wall** gets its strength from a coating of surface-bonding cement, which can also be tinted to match your home's exterior décor.

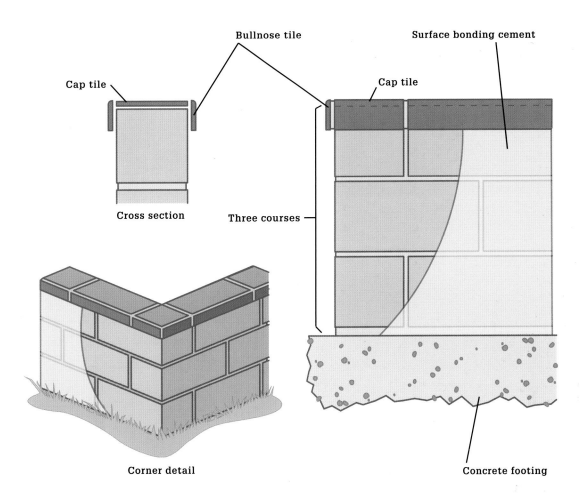

Bullnose tile

Cap tile

Surface bonding cement

Cap tile

Cross section

Three courses

Corner detail

Concrete footing

# How to Build a Mortarless Block Wall

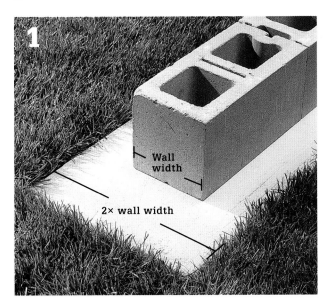

Wall width

2× wall width

**Complete a dry layout** of the first course on a concrete footing. Where less than half a block is needed, trim two blocks instead. For example, where three and one-third block lengths are required, use four blocks, and cut two of them to two-thirds their length. You'll end up with a stronger, more durable wall.

**Mark the corners** of the end blocks on the footing with a pencil. Then, remove the blocks and snap chalk lines to indicate where to lay the mortar bed and the initial course of block.

(continued)

**Mist the footing with water,** then lay a ¾"-thick bed of mortar on the footing. Take care to cover only the area inside the reference lines. The mortar must be firm enough to prevent the first course from sagging.

**Lay the first course,** starting at one end and placing blocks in the mortar bed with no spacing in between. Use solid-faced blocks on the ends of the wall and check the course for level. If your wall is longer than 20 ft., consider inclusion of an expansion joint.

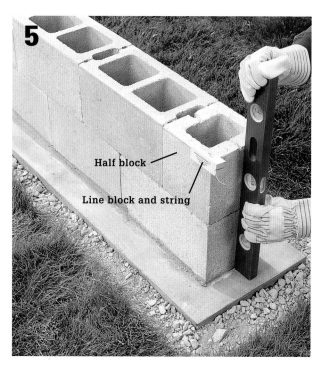

Half block

Line block and string

**Lay subsequent courses one at a time** using a level to check for plumb and line blocks to check for level. Begin courses with solid-faced blocks at each end. Use half blocks to establish a running bond pattern.

**Variation:** For walls with corners, begin the second course with a full-size end block that overlaps the joint between wall sections in the first course. Start the perpendicular run with a full stretcher block butted against the end block, as shown here. Repeat the alternation for each course.

**Lay wire mesh** over the next to last course. Install the top course, and then fill the block hollows with mortar, and trowel the surface smooth.

**Mix surface-bonding cement** thoroughly in a mortar box to achieve a firm, workable consistency. Eliminate all lumps during mixing. To color the cement, add coloring agent to the mixing water, or as directed by the manufacturer.

**Apply the cement** to the blocks after misting a workable section with water (to prevent premature drying). Working from the bottom up, spread the cement in an even ¼"-thick layer using a finishing trowel. Texture the surface, if desired.

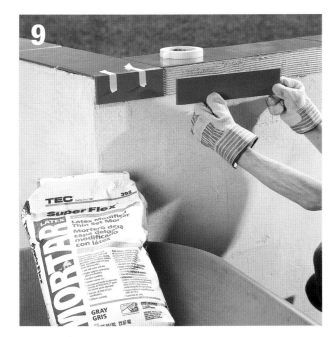

**Lay out the tiles on top of the wall.** Adjust the layout as needed so that cut tiles at the ends of the wall will be roughly the same size. Apply thinset mortar with a notched trowel, and set the tile, using tile spacers to set the grout joints. After the mortar sets on the top tile, install the bullnose tiles along both side edges, using tape to keep the tiles from slipping until the mortar sets.

**Grout the tiles after the mortar has fully cured.** Mix the grout as directed, and spread it over the tiles with a rubber grout float, packing it into the joints with the float held flat. Then, scrape off excess grout by dragging the float across the joints diagonally, with the float held at a 60° angle. Clean the tiles with a damp sponge. Apply sealer to the grout after it has cured completely.

# Poured Concrete Wall

Building vertically with poured concrete introduces a whole new dimension to this ever-versatile material. And as much as walls may seem more challenging than slabs or casting projects, the basic building process is just as simple and straightforward. Construct forms using ordinary materials, then fill them with concrete and finish the surface. While tall concrete walls and load-bearing structures require careful engineering and professional skills, a low partition wall for a patio or garden can be a great do-it-yourself project.

The first rule of concrete wall building is knowing that the entire job relies on the strength of the form. A cubic foot of concrete weighs about 140 pounds, which means that a three-foot-tall wall that is six inches thick weighs 210 pounds for each linear foot. If the wall is 10 feet long, the form must contain over a ton of wet concrete. And the taller the wall, the greater the pressure on the base of the form. If the form has a weak spot and the concrete breaks through (known as a blowout), there's little chance of saving the project. So be sure to brace, stake, and tie your form carefully.

This project shows you the basic steps for building a three-foot-high partition wall. This type of wall can typically be built on a poured concrete footing or a reinforced slab that's at least four inches thick. When planning your project, consult your local building department for specific requirements such as wall size, footing specifications, and metal reinforcement in the wall. *Note: This wall design is not suitable for retaining walls, tall walls, or load-bearing walls.*

For help with building a new footing, see pages 40 to 45. The footing should be at least 12 inches wide (twice the wall thickness) and at least 6 inches thick (equal to the wall thickness), and it must extend below the frost line (or in accordance with the local Building Code). If your wall will stand on a concrete patio or other slab, the sidebar on pages 164 and 165 shows you how to install rebar in the slab for anchoring the wall.

**In any setting,** a poured concrete wall offers clean, sleek lines and a reassuringly solid presence. You can leave the wall exposed to display its natural coloring and texture. For a custom design element, add color to the concrete mix or decorate any of the wall's surfaces with stucco, tile, or other masonry finishes.

# Wall Form Construction

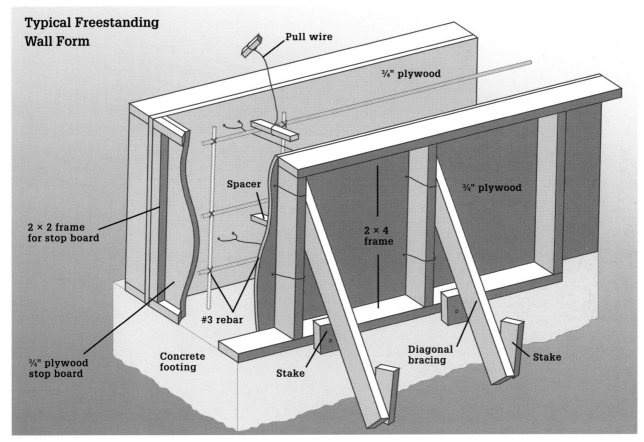

**Typical Freestanding Wall Form**

Pull wire

¾" plywood

Spacer

2 × 2 frame for stop board

¾" plywood

2 × 4 frame

#3 rebar

¾" plywood stop board

Concrete footing

Stake

Diagonal bracing

Stake

**A typical wall form** is built with two framed sides (much like a standard 2 × 4 stud wall) covered with ¾" plywood. The two sides are joined together at each end by a stop board, which also shapes the end of the finished wall. The form is braced and staked in position. Tie wires prevent the sides of the form from spreading under the force of the concrete. Temporary spacers maintain proper spacing between the sides while the form is empty; these are pulled out once the concrete is placed.

## Securing Braces on a Concrete Slab ▸

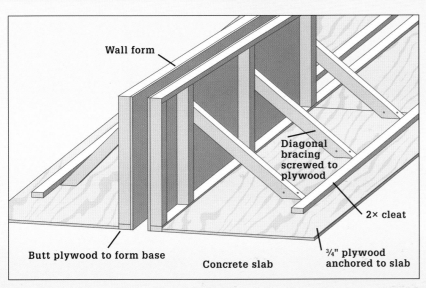

Wall form

Diagonal bracing screwed to plywood

2× cleat

Butt plywood to form base

Concrete slab

¾" plywood anchored to slab

**Fasten sheets** of ¾" plywood to the slab as an anchoring surface for form braces. Fasten the plywood with a few heavy-duty masonry screws driven into the slab. Butt the sheets against the bottom of the form to provide the same support you'd get from stakes. Screw diagonal form bracing directly to the plywood. You can also add a cleat behind the braces for extra support.

# How to Create a Poured Concrete Wall

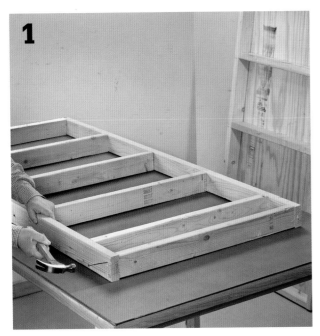

**Build the frames for the form sides** from 2 × 4 lumber and 16d nails. Include a stud at each end and every 16" in between. Plan an extra 2¼" of wall length for each stop board. For walls longer than 8 ft., build additional frames.

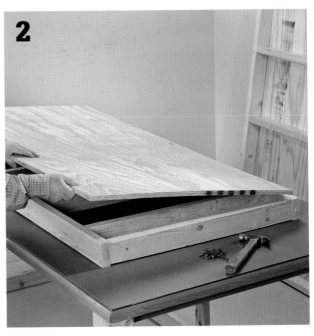

**Cut one piece of ¾" plywood** for each side frame. Fasten the plywood to frames with 8d nails driven through the plywood and into the framing. Make sure the top edges of the panels are straight and flush with the frames.

**Drill holes for the tie wires.** At each stud location, drill two pairs of ⅛" holes evenly spaced, and keep the holes close to the stud faces. Drill matching holes on the other form side.

Tie wire

Footing anchors

**Pour a concrete footing** (see pages 42 to 45) and set #3 rebar anchors into the concrete at 24" intervals. Once the footing has dried, cut #3 rebar for three horizontal runs, 4" shorter than the wall length. Tie the short pieces to the footing anchors using 8-gauge tie wire, then tie the horizontal pieces to the verticals, spacing them 12" apart and keeping their ends 2" from the wall ends.

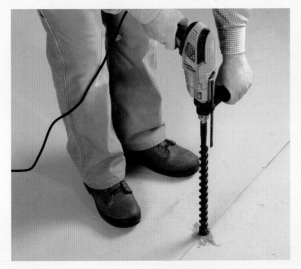

**A standard, reinforced 4"-thick concrete slab** can be a suitable foundation for a low partition wall like the one shown in this project. The slab must be in good condition, with no significant cracks or changes in level, and you should place the wall several inches away from the slab edge to ensure adequate support. To anchor the new wall to the slab and provide lateral stability, you'll need to install rebar anchors in the slab, following the basic steps shown here. But before going ahead with the project, be sure to have your plans approved by the local building department.

**Mark the locations for the rebar anchors** along the wall center: position an anchor 4" from each end of the wall and every 24" in between. At each location, drill a 1½"-diameter hole straight down into the concrete using a hammer drill and 1½" masonry bit (above, left). Make the holes 3" deep. Spray out the holes to remove all dust and debris using an air compressor with a trigger-type nozzle. Cut six pieces of #4 rebar at 16". Mix exterior-use anchoring cement to a pourable consistency. Insert the rods into the holes, then fill the hole with the cement (above, right). Hold the rods plumb until the cement sets (about 10 minutes). Let the cement cure for 24 hours.

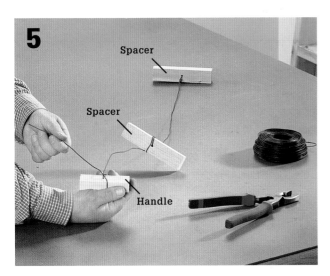

**5**

**Cut 1 × 2 spacers at 6",** one piece for each set of tie wire holes. These temporary spacers will be used to maintain the form width. Tie each pair of spacers to a pull wire, spacing them to match the hole spacing. Then attach a piece of scrap wood to the end of the pull wire to serve as a handle.

**6**

**Set the form sides in place.** Install the stop boards with 2 × 2 frames for backing; fasten the frames to the form sides with screws. Tie a loop of wire through each set of tie wire holes, and position a spacer near each loop. Use a stick or scrap of pipe to twist the loop strands together, pulling the form sides inward, tight against the spacers.

(continued)

**Make sure the form is centered over the footing.** Check that the sides are plumb and the top is level. Secure the form with stakes and braces: install a diagonal brace at each stud location, and stake along the bottom of the form sides every 12". Fasten all stakes and braces to the form framing with screws. For long walls, join additional side pieces with screws for a tight joint with no gapping along the plywood seam. Brace the studs directly behind the joint between sections. Coat the insides of the form with a release agent. If building on a slab (above, right), construct the form and then attach as a unit (see page 165).

**Mix the first batches of concrete,** being careful not to add too much water—a soupy mix results in weakened concrete.

**Place the concrete in the forms.** Start at the ends and work toward the center, filling the form about halfway up (no more than 20" deep). Rap on the forms to settle out air bubbles and then fill to the top. Remove the spacers as you proceed.

**Use a shovel** to stab into the concrete to work it around the rebar and eliminate air pockets. Continue to rap the sides of the forms with a hammer to help settle the concrete against the forms.

**Screed the top of the wall flat** with a 2 × 4, removing spacers as you work. After the bleed water disappears, float or trowel the top surface of the wall for the desired finish. Also round over the edges of the wall with an edger, if desired.

**Cover the wall with plastic** and let it cure for two or three days. Remove the plastic. Sprinkle with water on hot or dry days to keep concrete from drying too quickly.

**Cut the loops of tie wire** and remove the forms. Trim the tie wires below the surface of the concrete, and then patch the depressions with quick-setting cement or fast-set repair mortar. Trowel the patches flush with the wall surface. *Option: to achieve a consistent wall color and texture apply heavy duty masonry coating with acrylic fortifier using a masonry brush.*

# Interlocking Block Retaining Wall

Sloping areas of a yard may be fun for the kids to play on, but they can certainly limit your usable space for amenities like patios and gardens. When you need more flat ground or simply want to reshape nature's contours a bit, a low retaining wall is the answer. Retaining walls cut into a slope—and in some cases, replace the slope—bridging the upper and lower levels while adding more useable area to both.

Low retaining walls can be built with a variety of materials, including landscape timbers (pages 174 to 175), natural stone (pages 176 to 179), and poured concrete (pages 180 to 186). But by far the most popular material for do-it-yourself projects is interlocking concrete block made specifically for retaining walls. This block requires no mortar—most types are simply stacked in ordered rows—and it has flanges (or pins) that automatically set the batter for the wall (the backward lean that most retaining walls have for added strength). Interlocking block is available at home and garden centers and landscape suppliers. Most types have roughly textured faces to mimic the look of natural stone.

Due to the structural factors involved, the recommended height limit for do-it-yourself retaining walls is three feet. Anything higher is best left to a professional. As walls get taller, the physical stresses involved and resulting potential problems rise dramatically. Retaining walls of any size may be governed by the local Building Code; contact your city's building department to learn about construction specifications and permit requirements.

## Tools & Materials ▸

| | |
|---|---|
| Wheelbarrow | Caulk gun |
| Shovel | Circular saw with |
| Line level | masonry cutting blade |
| Hand tamper | Stakes |
| Tamping machine | Mason's string |
| (available | Landscape fabric |
| for rent) | Compactable gravel |
| Small maul | Perforated drain pipe |
| Masonry chisel | Coarse backfill material |
| Eye and | Construction adhesive |
| ear protection | Excavation tools |
| Work gloves | Interlocking block |
| 4-ft. level | Flour or marking paint |
| Tape measure | |

**Interlocking concrete block** is the only retaining wall material that comes ready to install. With little or no cutting, you can build a wall with straight lines, curves, or steps, or have it conform to a slope on one or both ends.

# Options for Positioning a Retaining Wall

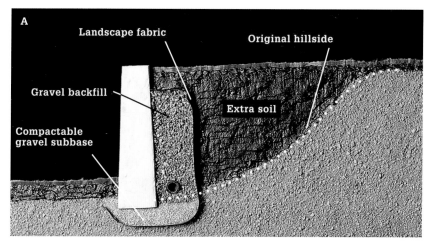

**A**
Landscape fabric
Original hillside
Gravel backfill
Extra soil
Compactable gravel subbase

**B**
Soil removed from base of hill
Original hillside
Compactable gravel subbase

**Structural features** for all retaining walls include: a compactable gravel subbase to make a solid footing for the wall, crushed stone backfill, a perforated drain pipe to improve drainage behind the wall, and landscape fabric to keep the loose soil from washing into and clogging the gravel backfill. There are a couple different ways you can position a retaining wall on your slope:

**(A)** Increase the level area above the wall by positioning the wall well forward from the top of the hill. Fill in behind the wall with extra soil, which is available from sand-and-gravel companies.

**(B)** Keep the basic shape of your yard by positioning the wall near the top of the hillside. Use the soil removed at the base of the hill to fill in near the top of the wall.

## Tips for Building Retaining Walls ▸

**Backfill with crushed stone** and install a perforated drain pipe about 6" above the bottom of the backfill. Vent the pipe to the side or bottom of the retaining wall, where runoff water can flow away from the hillside without causing erosion.

**Make a stepped trench** when the ends of a retaining wall must blend into an existing hillside. Retaining walls are often designed so the ends curve or turn back into the slope.

# How to Build a Retaining Wall Using Interlocking Block

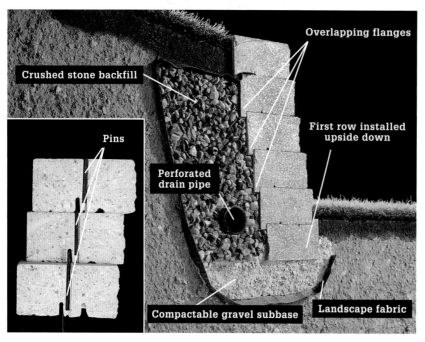

Crushed stone backfill

Overlapping flanges

Pins

First row installed upside down

Perforated drain pipe

Compactable gravel subbase

Landscape fabric

**1**

**Interlocking wall blocks do not need mortar.** Some types are held together with a system of overlapping flanges that automatically set the backward pitch (batter) as the blocks are stacked, as shown in this project. Other types of blocks use fiberglass pins (inset).

**Excavate the hillside, if necessary.** Allow 12" of space for crushed stone backfill between the back of the wall and the hillside. Use stakes to mark the front edge of the wall. Connect the stakes with mason's string, and use a line level to check for level.

**2**

**3**

**Dig out the bottom of the excavation** below ground level, so it is 6" lower than the height of the block. For example, if you use 6"-thick block, dig down 12". Measure down from the string in multiple spots to make sure the bottom base is level.

**Line the excavation with strips of landscape fabric** cut 3 ft. longer than the planned height of the wall. Make sure all seams overlap by at least 6".

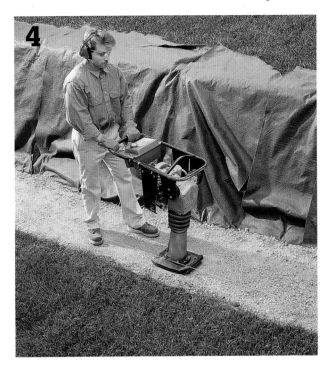

**Spread a 6" layer of compactable gravel** over the bottom of the excavation as a subbase and pack it thoroughly. A rented tamping machine, or jumping jack, works better than a hand tamper for packing the subbase.

**Lay the first course of block,** aligning the front edges with the mason's string. (When using flanged block, place the first course upside down and backward.) Check frequently with a level and adjust, if necessary, by adding or removing subbase material below the blocks.

**Lay the second course of block** according to manufacturer's instructions, checking to make sure the blocks are level. (Lay flanged block with the flanges tight against the underlying course.) Add 3 to 4" of gravel behind the block, and pack it with a hand tamper.

**Make half-blocks for the corners and ends of a wall,** and use them to stagger vertical joints between courses. Score full blocks with a circular saw and masonry blade, then break the blocks along the scored line with a maul and chisel.

(continued)

**Add and tamp crushed stone,** as needed, to create a slight downward pitch (about ¼" of height per foot of pipe) leading to the drain pipe outlet. Place the drain pipe on the crushed stone, 6" behind the wall, with the perforations face down. Make sure the pipe outlet is unobstructed. Lay courses of block until the wall is about 18" above ground level, staggering the vertical joints.

**Fill behind the wall with crushed stone,** and pack it thoroughly with the hand tamper. Lay the remaining courses of block, except for the cap row, backfilling with crushed stone and packing with the tamper as you go.

**Before laying the cap block,** fold the end of the landscape fabric over the crushed stone backfill. Add a thin layer of topsoil over the fabric, then pack it thoroughly with a hand tamper. Fold any excess landscape fabric back over the tamped soil.

**Apply construction adhesive to the top course of block,** then lay the cap block. Use topsoil to fill in behind the wall and to fill in the base at the front of the wall. Install sod or plants, as desired.

# How to Add a Curve to an Interlocking Block Retaining Wall

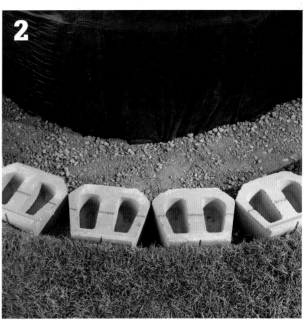

**1**

Right angle

**Outline the curve** by first driving a stake at each end and then driving another stake at the point where lines extended from the first stakes would form a right angle. Tie a mason's string to the right-angle stake, extended to match the distance to the other two stakes, establishing the radius of the curve. Mark the curve by swinging flour or spray paint at the string end, like a compass (see page 25).

**2**

**Excavate for the wall section,** following the curved layout line. To install the first course of landscape blocks, turn them upside down and backward and align them with the radius curve. Use a 4-ft. level to ensure the blocks sit level and are properly placed.

**3**

**4**

**Install subsequent courses** so the overlapping flange sits flush against the back of the blocks in the course below. As you install each course, the radius will change because of the backwards pitch of the wall, affecting the layout of the courses. Where necessary, trim blocks to size. Install using landscape construction adhesive, taking care to maintain the running bond.

**Use half blocks or cut blocks** to create finished ends on open ends of the wall.

# Timber Retaining Wall

## Tools & Materials ▸

Compactable gravel
Timber (5 × 6 or larger)
12" galvanized spikes
Eye and ear protection
Reciprocating saw
    and long wood blade

Excavation tools
Hand maul
Drill with bits
Landscape fabric
Hand tamper
Work gloves

A low retaining wall built with timbers follows many of the same construction steps as an interlocking block wall (see pages 168 to 173). All steps specific to timber construction are shown here.

When built properly, a timber retaining wall can have a life span of 15 to 20 years. Be sure to use pressure-treated lumber rated for ground contact, and build the wall with 5 × 6 or larger timbers; 4 × 4 and 4 × 6 sizes are not strong enough for retaining walls. Avoid using old, discarded railroad ties that have been soaked in creosote, which can leach into the soil and kill plants.

Cut the timbers with a reciprocating saw and long wood blade (or a chain saw, if you prefer). Before building the retaining wall, prepare the site as directed in steps one to three on page 170.

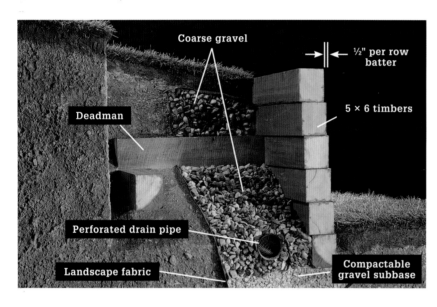

Coarse gravel

½" per row batter

5 × 6 timbers

Deadman

Perforated drain pipe

Landscape fabric

Compactable gravel subbase

**Timber retaining walls** must be anchored with "deadmen" that extend from the wall back into the soil. Deadmen prevent the wall from sagging under the weight of the soil. For best results with timber retaining walls, create a backward angle (batter) by setting each row of timbers ½" behind the preceding row. The first row of timbers should be buried.

## Tips for Strengthening a Timber Retaining Wall ▸

Timber depth equal to ½ exposed height

**Install vertical anchor posts** to reinforce the wall. Space the posts 3 ft. apart, and install them so the buried depth of each post is at least half the exposed height of the wall. Anchor posts are essential if it is not practical to install deadmen.

# How to Build a Retaining Wall Using Timbers

**Spread a 6"-layer of compactable gravel subbase** into the prepared trench, then tamp the subbase and begin laying timbers, following the same techniques as with interlocking blocks (steps 4 to 11, pages 171 to 172). Each row of timbers should be set with a ½" batter, and end joints should be staggered so they do not align.

**Use 12" galvanized spikes or reinforcement bars** to anchor the ends of each timber to the underlying timbers. Stagger the ends of the timbers to form strong corner joints. Drive additional spikes along the length of the timbers at 2-ft intervals. If you have trouble driving the spikes, drill pilot holes.

**Install deadmen,** spaced 4 ft. apart, midway up the wall. Build the deadmen by joining 3-ft-long lengths of timber with 12" spikes, then insert the ends through holes cut in the landscape fabric. Anchor deadmen to the wall with spikes. Install the remaining rows of timbers, and finish backfilling behind the wall (steps 6 to 11, pages 171 to 172).

**Improve drainage** by drilling weep holes through the second row of landscape timbers and into the gravel backfill using a spade bit. Space the holes 4 ft. apart, and angle them upward.

# Stone Retaining Wall

Rough-cut wall stones may be dry stacked (without mortar) into retaining walls, garden walls and other stonescape features. Dry-stack walls move and shift with the frost, and they drain well so they don't require deep footings and drain tiles.

In the project featured here, we use rough-split limestone blocks about eight inches by about four inches thick and in varying lengths. Walls like this may be built up to three feet tall, but keep them shorter if you can, to be safe. Building multiple short walls is often a more effective way to manage a slope than to build one taller wall. Called terracing, this practice requires some planning. Ideally, the flat ground between pairs of walls will be approximately the same size.

A dry-laid natural stone retaining wall is a very organic-looking structure compared to interlocking block retaining walls (pages 168 to 173). One way to exploit the natural look is to plant some of your favorite stone-garden perennials in the joints as you build the wall. Usually one plant or a cluster of three will add interest to a wall without suffocating it in vegetation or compromising its stability. Avoid plants that get very large or develop thick, woody roots or stems that may affect the stability of the wall.

A well-built retaining wall has a slight lean, called a batter, back into the slope. It has a solid base of compacted gravel, and the first course is set below grade for stability.

## Tools & Materials ▸

Mason's string
Line level
Stakes
Hand maul
Torpedo level
Straight 2 × 4
Hand tamper
Compactable gravel
Ashlar wall stone
Landscape fabric
Caulk gun
Block and
   stone adhesive
Excavation tools

Coarse sand
Drainage gravel
   (1½ to 3"
   river rock is
   recommended)
Stone chisel
4-ft. level
Tape measure
Hammer
Scissors
Work gloves
Eye and
   ear protection

**A natural stone retaining wall** blends into its surroundings immediately and only looks better with age. Building the wall with ashlar, or cut wall stone, is a much easier project than a wall built with round fieldstones or large boulders.

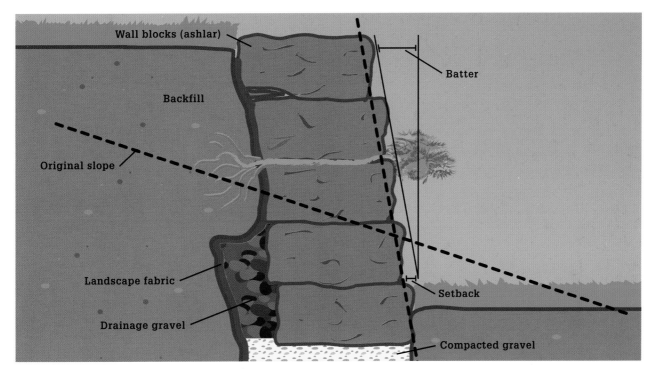

Our single-row retaining wall has a ½" batter, created by setting each course of stone ½" back from the face of the course below. The base of the wall includes a compacted gravel subbase topped with sand to help level the first course of stones. Roots of plants sewn into the wall crevices (an optional decorative embellishment) will eventually reach into the soil behind the wall.

## How to Build a Stone Retaining Wall

**Begin excavating the wall site.** Dig a trench for the base of wall, making it 6" wider than the wall thickness. If necessary, dig into the slope, creating a backward angle that roughly follows the ½" batter the wall will have. If desired, dig returns back into the slope at the end(s) of the wall.

**Measure the depth of the trench** against a level mason's string running parallel to the trench. The bottom of the trench should be level and 8" below grade (ground level) for the main section of wall and any returns. If the trench becomes too shallow due to natural contours, step it down the height of one stone.

(continued)

**Complete the wall base** by tamping the soil in the trench, and then adding a 3"-layer of compactable gravel and tamping it flat and level. Cover the gravel with landscape fabric, draping the fabric back over the slope. Add a 1"-layer of sand over the fabric in the trench area. Smooth and level the sand with a short 2 × 4 screed board, checking for level with a torpedo level set on the board.

**Set the first course with heavy stones,** laying long, square-ended stones at the corners first. *Tip: Organize your stones by size, and plan to set each course with stones of similar thicknesses.* Set up a level mason's string just in front of the top front edge of the course, letting the stones roughly guide the string placement.

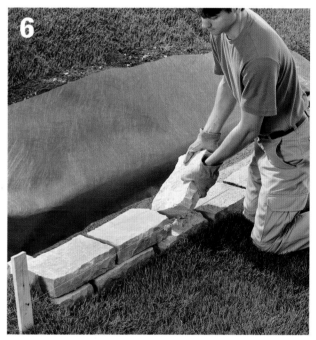

**Add or remove sand beneath the stones as needed** so they are nearly touching the string. Level the stones front to back with a torpedo level and side to side with a 4-ft. level. If necessary, use a hand maul and stone chisel to chip off irregularities from the edges of stones to improve their fit.

**Begin the second course,** starting with both ends of the wall face. Reset and level the mason's string at the height of the second course. Place the second-course stones back ½" from the front edges of the first-course stones, overlapping all joints of the first course to create a bond pattern.

**Shim beneath stones as needed** to level them or add stability, using stone shards and chips. Complete the second course over the main part of the wall.

**Complete the returns,** as applicable, maintaining the offset joint pattern with the first course. You may need to dig into the slope to create a level base for the return stones. Add a layer of compacted gravel under each return stone before setting it. Complete the remaining courses up to the final (capstone) course.

**Backfill behind the base of the wall** with drainage rock (not compactable gravel). For a low wall like this, 6 to 10" of gravel is usually sufficient; taller walls may require more gravel and possibly a drainage pipe. Pack the gravel down with a 2 × 4 to help it settle.

**Fold the landscape fabric over the gravel,** and backfill over the fabric with soil. (The fabric is there to prevent the soil from migrating into the gravel and out through the wall stones.) Trim the fabric just behind the back of the wall, near the top.

**Install the final course** using long, flat cap stones. Glue the caps in place with block and stone adhesive. After the glue dries, add soil behind the wall to the desired elevation for planting.

# Poured Concrete Retaining Wall

Poured concrete has advantages and disadvantages as a building material for structural garden walls, such as this retaining wall. On the plus side: it can conform to just about any size and shape you desire (within specific structural limitations); depending on your source, concrete can be a relatively inexpensive material; poured concrete is very longlasting; with professional engineering you can build higher with poured concrete than with most other wall materials. But if you live in a region where freeze/thaw cycles are problematic, you'll need to dig down deep (at least a foot past the frostline) and provide plenty of good drainage to keep your wall from developing vertical cracks.

A properly engineered retaining wall is designed using fairly complicated dimensional and force ratios. If the wall will be three feet or taller, you should have it engineered by a professional. Shorter retaining walls, sometimes called curb walls, often require less stringent engineering, especially if they are located in a garden setting or are to be used for planting beds or terracing. The wall seen here is built in a fairly cold climate, but the fact that the top is less than 36 inches above ground allows for a drainage base that is above the frostline, with the understanding that some shifting is likely to occur. The project was built in conjunction with poured concrete steps. Because the steps and walls are isolated with an isolation membrane they are regarded as independent structures and neither is required to have footings that extend below the frostline.

## Tools & Materials ▸

| | |
|---|---|
| Shovel | Landscape fabric |
| Circular saw | Drainage gravel |
| Power miter saw | Excavation tools |
| Table saw | Compactable gravel |
| Drill/driver | Eye and |
| Level | ear protection |
| Tamper or plate | Mason's string |
| compactor | Metal stakes |
| Bow rake | Hand maul |
| Wheelbarrow | Tie wire |
| Float | Construction |
| Edging tool | adhesive |
| ¾" plywood | Caulk gun |
| Lumber | Concrete vibrator |
| (2 × 4, 1 × 4) | (available |
| Deck screws | for rent) |
| Concrete | Rubber mallet |
| release agent | Angle iron |
| #3 rebar | Square tubing |
| Concrete | Stapler |
| Sheet plastic | Staples |
| Concrete colorant | Work gloves |
| (optional) | Magnesium trowel |
| 2½"-dia. ABS | or darby |
| plastic pipe | |

After

Before

**A poured concrete retaining wall** adds structure and permanence to your landscape. Build decorative forms and add coloring agents to the concrete to add decorative flair to this utilitarian structure.

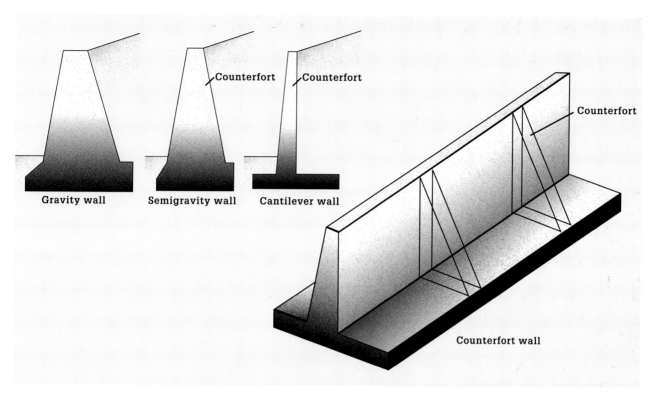

Gravity wall

Semigravity wall

Cantilever wall

Counterfort

Counterfort

Counterfort

Counterfort wall

**Poured concrete retaining** walls employ differing strategies to keep the earth at bay. Some, called gravity walls, rely almost exclusively on sheer mass to hold back the ground swell. These are very wide at the bottom and taper upward in both the front and the back. Unless you feel like pouring enough concrete to build a dam, don't plan on a gravity wall that's more than 3 ft. tall. A semigravity wall is somewhat sleeker than a gravity wall and employs internal reinforcement to help maintain its shape. A cantilevered wall has an integral bottom flange that extends back into the hillside where it is held down by the weight of the dirt that is backfilled on top of it. This helps keep the wall in place. A counterfort wall is a cantilevered wall that has diagonal reinforcements between the back face of the wall and the flange.

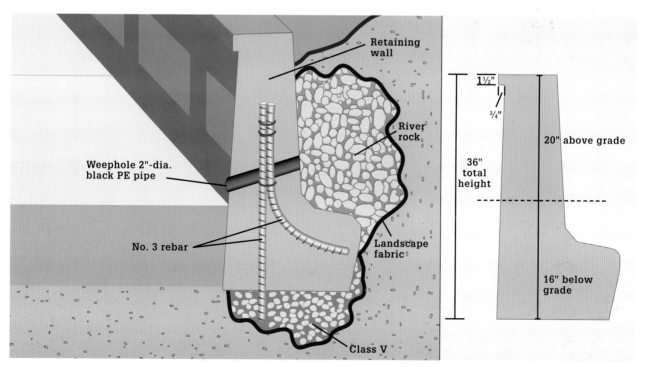

Retaining wall

River rock

Weephole 2"-dia. black PE pipe

No. 3 rebar

Landscape fabric

Class V

1½"

¾"

36" total height

20" above grade

16" below grade

**A cross-section** of the concrete retaining wall built here reveals internal re-bar reinforcement, and 2"-dia. pipe that's cast into the wall to allow drainage slightly above the grade.

# How to Build a Poured Concrete Retaining Wall

**Excavate the construction area** well beyond the edges of the planned wall. Reserve some soil for backfilling and transport some to lower areas in your yard that need building-up. For larger walls, you can save a lot of work by renting power equipment, or hiring an excavator. If your utilities company has flagged any pipes in the area you must dig around them using hand tools.

**Add a thick layer** (at least 4 to 6") of compactable gravel to the bottom of the excavation and tamp it thoroughly with a hand tamper or a rented plate compactor to create a solid foundation for the wall. Add additional base material in sandy or marshy soil.

— Straightedge cutting guide

**Cut the form boards,** usually from ¾"-thick exterior-grade plywood with one sanded face. You may also use dimensional lumber. Do not use oriented-strand board or particleboard because they have insufficient sheer strength. Do not use any sheet goods that can weaken and delaminate from exposure to wet concrete. Use a circular saw and cutting guide or a table saw to cut panel to width

Stake

Braces

**Level and stake the forms** after doing as much pre-assembly as you can, including attaching the 1 × 4 front forms to create the lip. Run mason's lines at the tops of the forms to use as a reference. Drive 2 × 4 stakes into the ground next to the form braces and attach the braces to the stakes with deck screws. Wherever possible, stake the forms by driving screws through the outer member so they can be removed to facilitate stripping off the forms.

**Stake the back forms** as well as the front forms. You'll have to get a little creative for this task in many cases, because much depends on the condition of the soil or ground surrounding the wall, as well as access to the forms both during and after the pour. Here, strips of plywood are secured to metal stakes driven into the hill behind the forms. The strips are then screwed to the braces on the back wall forms to hold them in position.

**Add re-bar reinforcement** to tie the integral footing and the wall together. Drive a length of rebar into the bottom of the wall area and then bend another piece and attach it to the rebar stake with wire. Install these reinforcements every 2 to 3 ft. For extra strength, connect them with a horizontal piece of rebar.

**Install weep holes.** Choose drain pipe (black ABS plastic is a good choice) around 2½" in diameter and cut lengths that are equal to the distance between the inside faces of the plywood form boards. Insert the pipes into the bottoms of the form so each end is flush against the inside face of the form. Install a weep hole drain every 6 ft. or so. Drive a long deck screw through the front panel and the back panel so the screw penetrates the form inside the weep drain, near the top. The ends of the screws will create supports for the drain pipe if the friction fails during the pour (as is likely).

## Add Decorative Elements to Forms ▸

One of the beauties of poured concrete is that it is pliable enough to conform to an endless number of decorative forming schemes. For the retaining wall seen here, a series of particleboard panels with beveled edges are attached at regular intervals to the inside faces of the outer form to create a very familiar recessed panel appearance. The panels (beveled edges are cut on a table saw) should be attached with construction adhesive and short screws so there are not gaps that concrete can seep into. Be sure to use a release agent.

Vegetable oil

**Apply a release agent** to the inside surfaces of the forms. You can purchase commercial release agent product, or simply brush on a light coat of vegetable oil or mineral oil. Do not use soap (it can weaken the concrete surface) or motor oil (it is a pollutant).

(continued)

**9**

**Place the concrete.** For most walls, 4000 psi mixture (sometimes called 5-bag) is called for. You can mix concrete by hand, rent an on-site power mixer, haul your own premixed concrete in a rental trailer, or have the concrete delivered. Begin filling the form at one end of the wall and work your way methodically toward the other end. Have plenty of help so you can start tooling the concrete as soon as possible.

## Colored Concrete ▶

The concrete mix seen here was pre-tinted at the concrete mixing plant. If you do not want a gray concrete structure, using tinted concrete adds color without the need to refresh paint or stain. However, the process is not cheap (an additional $60 per yard), the final color is unpredictable, and you'll have to tint the concrete to match if you need to repair the structure in the future. The pigment that is added can also have unforeseen effects on the concrete mixture, such as accelerating the set-up time.

**10**

**Use a splash guard** to direct the concrete out of the chute and into the form without spilling. The concrete supplier should have hoops, shovels and other tools to assist with form filling.

**11**

Concrete vibrator

**Settle the concrete** in the forms as you work. For best results, rent a concrete vibrator and vibrate thoroughly before screeding. Do not get carried away—over-vibrating the concrete can cause the ingredients to separate. A less effective alternative for vibrating (requiring no rental tools) is to work the concrete in with a shovel and settle it by rapping the forms with a rubber mallet.

**Strike off, or "screed," the concrete** so it is level with the tops of the forms. Use a piece of angle iron attached to square tubing, or a 2 × 4, as a screed. Move the screed slowly across the forms in a sawing motion. Do not get ahead of the concrete. The material behind the screed should be smooth and level, with no dips or voids.

**Tool the concrete** once the bleed water evaporates, if desired. For a smoother top, float the surface with a magnesium trowel or darby. Run an edger along the top edges on at least the front edge, and preferably the back as well.

**Cover the concrete surface** with plastic sheeting to cure for at least 48 hours, especially during hot weather. If it is very hot and dry, lift the plastic off and douse the concrete with fresh water twice a day to slow down the drying. Drying too fast can cause cracking and other concrete failures. Wait at least two days and remove the forms.

**To backfill,** first shovel in an 8 to 12" deep layer of drainage gravel (1 to 2" dia.), then place a layer of landscape fabric over the gravel to keep weeds and other plants out. Shovel dirt over the fabric and tamp it lightly until the desired grade is achieved.

# Repairs for Walls & Fences

Nature takes its toll on all outdoor structures. Wood fences are the most vulnerable, being continuously tested by water and sunlight. Stone walls, while extremely durable, can be prone to shifting and settling soil and to improper stacking during construction. Brick structures wear most in their mortar joints, which are softer than the brick and inevitably deteriorate over time.

The repair projects in this chapter cover these most common problems for stone and brick walls and wood fences. Other types of fencing require much less maintenance than wood. In fact, if your vinyl, composite, or metal fence does need repairs (apart from basics like loose or missing fasteners), you'll probably have to replace the damaged parts with new ones and should contact the manufacturer for recommendations (or to discuss your warranty coverage).

It's important to note that structural problems with all walls and fences should not be ignored. Even short masonry walls can topple under the right conditions, and failing fences can easily blow over in a strong wind. Another thing to consider is the health of the structure as a whole. For example, if one fence post is rotted at its base, the two neighboring posts must shoulder the added burden, leaving them more vulnerable to problems and possible failure.

### In this chapter:
- Stone Walls
- Brick Structures
- Wood Fences

# Stone Walls

Damage to stonework is typically caused by frost heave, erosion or deterioration of mortar, or by stones that have worked out of place. Dry-stone walls are more susceptible to erosion and popping, while mortared walls develop cracks that admit water, which can freeze and cause further damage.

Inspect stone structures once a year for signs of damage and deterioration. Replacing a stone or repointing crumbling mortar now will save you work in the long run.

A leaning stone wall probably suffers from erosion or foundation problems, and can be dangerous if neglected. If you have the time, you can tear down and rebuild dry-laid structures, but mortared structures with excessive lean need professional help.

## Tools & Materials ›

| | |
|---|---|
| Chalk | Trowels for mixing and pointing |
| Maul | Stiff-bristle brush |
| Chisel | Carpet-covered 2 × 4 |
| Camera | Compactable gravel |
| Shovel | Replacement stones |
| Hand tamper | Type M mortar |
| Level | Mortar tint |
| Batter gauge | Screwdriver bent in a vice |
| Mortar bag | Spray bottle |
| Masonry chisels | Eye and ear protection |
| Wood shims | Work gloves |

**Stones in a wall can become dislodged** due to soil settling, erosion, or seasonal freeze-thaw cycles. Make the necessary repairs before the problem migrates to other areas.

## Replacing Popped Stones ›

**Return a popped stone** to its original position. If other stones have settled in its place, drive shims between neighboring stones to make room for the popped stone. Be careful not to wedge too far.

**Use a 2 × 4 covered with carpet** to avoid damaging the stone when hammering it into place. After hammering, make sure a replacement stone hasn't damaged or dislodged the adjoining stones. Remove shims.

# How to Rebuild a Dry Stone Wall Section

**Before you start,** study the wall and determine how much of it needs to be rebuilt. Plan to dismantle the wall in a "V" shape, centered on the damaged section. Number each stone and mark its orientation with chalk so you can rebuild it following the original design. *Tip: Photograph the wall, making sure the markings are visible.*

**Cap stones are often set in a mortar bed** atop the last course of stone. You may need to chip out the mortar with a maul and chisel to remove the cap stones. Remove the marked stones, taking care to check the overall stability of the wall as you work.

**Rebuild the wall** one course at a time, using replacement stones only when necessary. Start each course at the ends and work toward the center. On thick walls, set the face stones first, then fill in the center with smaller stones. Check your work with a level, and use a batter gauge to maintain the batter of the wall. If your capstones were mortared, re-lay them in fresh mortar. Wash off the chalk with water and a stiff-bristle brush.

## Improve Drainage ▶

**If you're rebuilding because of erosion,** dig a trench at least 6" deep under the damaged area, and fill it with compactable gravel. Tamp the gravel with a hand tamper. This will improve drainage and prevent water from washing soil out from beneath the wall.

**Tint mortar for repair work** so it blends with the existing mortar. Mix several samples of mortar, adding a different amount of tint to each, and allow them to dry thoroughly. Compare each sample to the old mortar, and choose the closest match.

**Use a mortar bag** to restore weathered and damaged mortar joints over an entire structure. Remove loose mortar (see below) and clean all surfaces with a stiff-bristle brush and water. Dampen the joints before tuck-pointing, and cover all of the joints, smoothing and brushing as necessary.

## How to Repoint Mortar Joints

**Carefully rake out cracked and crumbling mortar,** stopping when you reach solid mortar. Remove loose mortar and debris with a stiff-bristle brush. *Tip: Rake the joints with a chisel and maul, or make your own raking tool by placing an old screwdriver in a vice and bending the shaft about 45°.*

**Mix Type M mortar,** then dampen the repair surfaces with clean water. Working from the top down, pack mortar into the crevices using a pointing trowel. Smooth the mortar when it has set up enough to resist light finger pressure. Remove excess mortar with a stiff-bristle brush.

# How to Replace a Stone in a Mortared Wall

**Remove the damaged stone** by chiseling out the surrounding mortar using a masonry chisel or a modified screwdriver (opposite page). Drive the chisel toward the damaged stone to avoid harming neighboring stones. Once the stone is out, chisel the surfaces inside the cavity as smooth as possible.

**Brush out the cavity** to remove loose mortar and debris. Test the surrounding mortar, and chisel or scrape out any mortar that isn't firmly bonded.

**Dry-fit the replacement stone.** The stone should be stable in the cavity and blend with the rest of the wall. You can mark the stone with chalk and cut it to fit (pages 32 to 33), but excessive cutting will result in a conspicuous repair.

**Mist the stone and cavity lightly,** then apply Type M mortar around the inside of the cavity using a trowel. Butter all mating sides of the replacement stone. Insert the stone and wiggle it forcefully to remove any air pockets. Use a pointing trowel to pack the mortar solidly around the stone. Smooth the mortar when it has set up.

# Brick Structures

The most common repair for outdoor brick structures is tuck-pointing, the process of filling old, failing mortar joints with fresh mortar. Tuck-pointing is a time-consuming and somewhat tedious job, but it isn't difficult, and it's the best way to restore as much of the original look and strength of the wall as possible. For minor cracking and limited mortar failure, you can also make repairs with a sanded mortar repair caulk, which has coloring and a rough texture that mimic that of real mortar.

Other brick repairs include replacing a damaged brick and rebuilding a small section of a brick structure. These are generally safe repairs, provided the damage is localized and the wall is not tall or load-bearing. For extensive damage to any type of brick structure, consult a professional before attempting any repairs.

## Concrete Fortifier ▸

**Add acrylic- or latex-based concrete fortifier** to mortar for making repairs. This increases the mortar's strength and improves its ability to bond. For best results, tint the new mortar to match the original (see page 190). Mortar repairs can be highly conspicuous when the coloring of the new mortar is off.

## Tools & Materials ▸

| | | | |
|---|---|---|---|
| Raking tool | Bricklayer's hammer | Stiff-bristle brush | Replacement bricks |
| Mortar hawk | Mason's or stone chisel | Mortar | Eye and ear protection |
| Tuck-pointer | Pointing trowel | Concrete fortifier | Work gloves |
| Jointing tool | Drill with masonry disc | | |

Before

After

**Make timely repairs to brick structures.** Tuck-pointing deteriorated mortar joints is a common repair that, like other masonry fixes, improves the appearance of the structure or surface and helps prevent further damage.

# How to Tuck-point Mortar Joints

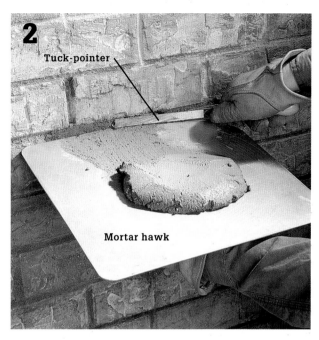

**Clean out loose or deteriorated mortar** to a depth of ¼ to ¾". Use a mortar raking tool (top) first, then switch to a masonry chisel and a hammer (bottom) if the mortar is stubborn. Clear away all loose debris, and dampen the surface with water before applying fresh mortar.

**Mix the mortar,** adding concrete fortifier; add tint if necessary. Load mortar onto a mortar hawk, then push it into the horizontal joints with a tuck-pointer. Apply mortar in ¼"-thick layers, and let each layer dry for 30 minutes before applying another. Fill the joints until the mortar is flush with the face of the brick or block.

**Apply the first layer of mortar** into the vertical joints by scooping mortar onto the back of a tuck-pointer, and pressing it into the joint. Work from the top downward.

**After the final layer of mortar is applied,** smooth the joints with a jointing tool that matches the profile of the old mortar joints. Tool the horizontal joints first. Let the mortar dry until it is crumbly, then brush off the excess mortar with a stiff-bristle brush.

# How to Replace a Damaged Brick

**Score the damaged brick** so it will break apart more easily for removal: use a drill with a masonry-cutting disc to score lines along the surface of the brick and in the mortar joints surrounding the brick.

**Use a mason's chisel and hammer** to break apart the damaged brick along the scored lines. Rap sharply on the chisel with the hammer, being careful not to damage surrounding bricks. *Tip: Save fragments to use as a color reference when you shop for replacement bricks.*

**Chisel out any remaining mortar in the cavity,** then brush out debris with a stiff-bristle or wire brush to create a clean surface for the new mortar. Rinse the surface of the repair area with water.

**Mix the mortar for the repair,** adding concrete fortifier to the mixture, and tint if needed to match old mortar. Use a pointing trowel to apply a 1"-thick layer of mortar at the bottom and sides of the cavity.

**5** Dampen the replacement brick slightly, then apply mortar to the ends and top of the brick. Fit the brick into the cavity and rap it with the handle of the trowel until the face is flush with the surrounding bricks. If needed, press additional mortar into the joints with a pointing trowel.

**6** Scrape away excess mortar with a masonry trowel, then smooth the joints with a jointing tool that matches the profile of the surrounding mortar joints. Let the mortar set until crumbly, then brush the joints to remove excess mortar.

## Tips for Removing & Replacing Several Bricks ▸

**For walls with extensive damage,** remove bricks from the top down, one row at a time, until the entire damaged area is removed. Replace bricks using the techniques shown above and on pages 154 to 157. *Caution: do not dismantle load-bearing brick structures like foundation walls—consult a professional mason for these repairs.*

**For walls with internal damaged areas,** remove only the damaged section, keeping the upper layers intact if they are in good condition. Do not remove more than four adjacent bricks in one area—if the damaged area is larger, it will require temporary support, which is a job best left to a professional mason.

# Wood Fences

Most wood fences are pretty simple. Unfortunately, the bad effects of exposure to the elements take a toll on wood fences. Posts will rot where they meet the ground, as will bottom stringers that sometimes sit in snow for long periods and are splashed with rain water the rest of the year. Older fences also have a tendency to lean one way or the other, usually as a result of frost movement in the ground. And no matter how good the structural condition of your fence, if it's been painted once it will need to be painted again and again for the rest of its life. This section will help you to deal with problems like these.

You can replace rotted posts, but it's a lot of work and it's not really necessary. After all, the problem with the post is where it meets the ground. The part that holds the stringers and boards is usually fine. So, it makes better sense to reinforce the bottom of the post with a stub post installed right next to it.

Sometimes fence stringers are just loose and they can be tightened by replacing the fasteners that hold them to the posts. Galvanized nails or screws are good for this job. If replacing the fasteners doesn't stiffen the stringer enough, add galvanized T-brackets to the joint to provide extra support. But occasionally a stringer is so rotted that it has to be replaced. This job can be a pain but if approached methodically should take only an hour or two.

Thanks to gravity, problems with wood gates usually involve loose hinges or hinge mounting screws, sagging gates, and leaning gate posts. Fortunately, all of these maladies are easy to remedy, often without replacing any parts.

## Specialized Tools ▶

**Come-along tool**—A hand-operated winch, typically with a steel hook on one end and a cable-mounted steel hook on the other. Made to allow one person to move heavy loads, or to tighten shipping straps.

**Turnbuckle cable**—A galvanized hardware fixture that has a right-hand threaded screw eye on one end and a left-hand screw eye on the other, both of which are connected to galvanized steel cables. By turning the turnbuckle the wires are tightened or loosened.

**Fence damage often looks worse than it is.** You could undertake a back-breaking demolition and rebuild to deal with a rotted post...but there is an easier way.

# Tools & Materials for Fence and Gate Repairs*

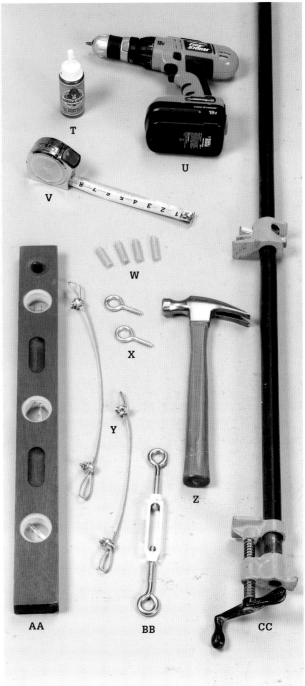

**Pictured:** (A) dry concrete mix, (B) primer and paint, (C) caulk, (D) come-along tool, (E) random orbit sander, (F) cold chisel, (G) drill, (H) hardware, (I) sandpaper, (J) nailset, (K) maul, (L) flat pry bar, (M) paint scraper, (N) hammer, (O) reciprocating saw, (P) level, (Q) shovel, (R) 4-ft. length of 1½" steel pipe, (S) bar clamps, (T) waterproof glue, (U) drill and bits, (V) tape measure, (W) ½ × 2" hardwood dowels, (X) screw eye, (Y) turnbuckle cables, (Z) hammer, (AA) level, (BB) turnbuckles, (CC) pipe clamp. *Not pictured: lumber (2 × 4, 4 × 4); screwdriver; putty knife or painter's 5-in-1 tool; detail sander; caulk gun; paint brush; eye and ear protection; protective mask; bits; rag.

# How to Fix a Rotted Post and Stringer

**To repair a rotted post,** first break up the concrete collar (if the post has one) and cut off the rotted section at the bottom of the post with a hand saw or reciprocating saw. Then dig out the cut section of the old post and collar with a shovel.

**Brace the fence to compensate for the cut post,** then cut a stub post to length and put it in the hole next to the old post. Plumb it and brace it in the plumb position. Then fill around the stub with concrete.

**Make sure the fence stringers are level** and the fence boards are vertical. Drill countersunk guide holes all the way through the stub post, the old post and the fence stringers and fence boards. Tap a carriage bolt into each hole. Add and tighten washers and nuts.

**Tighten a loose stringer** by replacing its fasteners with galvanized nails or screws. Be sure to set the nail heads below the surface with a nailset.

**To improve the strength of a stringer-to-post joint,** install a galvanized T-bracket with galvanized or stainless steel screws.

**To replace a rotted stringer,** first remove the fasteners from the face of the fence boards. If screws were used just back them out with a cordless drill. But if you find nails, the fence boards have to be pried away from the stringer with a flat pry bar. Once the nails are loose, pull them with a hammer.

**Cut a new stringer to length** and slide it between the posts. Support the stringer on a block of wood clamped to the post on both ends.

**Attach the new stringer to the posts** by driving angled galvanized screws through the stringer and into the post. Once the stringers are stable, screw the fence boards to the stringer.

# How to Straighten a Leaning Fence

**To fix a leaning fence,** first dig around the base of all the leaning posts to free them for movement. Then push against the fence with several friends until it is plumb. When it is, brace it in place.

**If you are working alone,** you can straighten the fence using a come-along tool. To use it, first drive a screw eye near the top of any leaning post.

**Next, drive a length of steel pipe into the ground,** at a 45° angle, about 5 ft. from the post. Hook the come-along to the screw eye and the pipe and start ratcheting the come-along. This will pull over the top of the fence until it's plumb. Brace the first post and move on to others that are leaning.

**Once all the posts have been pulled straight,** recheck the fence for plumb, then fill around the bottoms of the posts with concrete. Wait at least 3 days for the concrete to dry, then remove the braces.

# How to Paint a Fence

**Scrape off loose and flaking paint from all surfaces** using a paint scraper. You can use smaller tools, such as a putty knife or painter's 5-in-1 tool, for fitting into corners and tight spaces. If you find any loose fasteners, re-drive or replace them.

**Sand the scraped areas** to smooth out the edges between bare and painted wood using 80-grit sandpaper. If necessary, sand other painted surfaces with 100-grit paper to ensure a good bond with the new paint. *Tip: An orbital sander works best for large areas, and a detail sander (shown here) is great for tight spots.*

**Prime and paint the fence.** Start by applying a thorough coat of exterior primer. When the primer dries, fill any holes and cracks with paintable exterior caulk. Paint the fence with 2 top coats of quality exterior paint (either oil-based or 100% acrylic), waiting a day between coats.

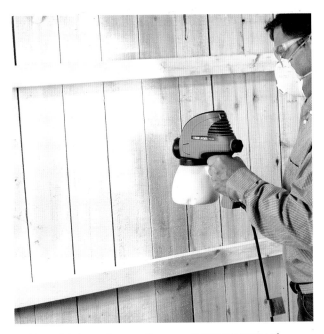

**Option:** Paint your fence with a power sprayer. Many of today's DIY sprayers are ideal for fences and other jobs that have painted surfaces. You can choose from corded and cordless models. Most sprayers feed from a cup attached to the gun; many corded models can also feed directly from a paint can or bucket.

# How to Fix a Worn Gate

**If the hinge screws have loosened,** remove the gate and drill ½"-diameter holes just under 2" deep in the posts at each hinge screw hole. A spade bit works well for this job.

**Coat ½ × 2" dowels** with exterior-rated glue and drive them into the holes with a hammer. Wipe up any glue squeeze-out with a rag.

**Brace the gate in the opening** and mark the exact location of the hinge screws. Then bore pilot holes and drive the screws flush to the surface of the hinge.

**If your gate is out of square,** use a pipe clamp to force it back into a square shape. Take diagonal measurements of the frame and apply the clamp on the diagonal that has the longer measurement. Tighten the clamp screw until both diagonal measurements are the same.

**Screw a temporary brace** across one corner of the gate. This should keep the gate square when you remove the pipe clamp.

**Drive screw eyes** into the top outside and bottom inside corners of the gate. Then install a turnbuckle cable between the two. Tighten the turnbuckle, remove the temporary brace and re-hang the gate.

## Fixing a Leaning Gate Post ▸

If your gate post leans, plumb the gate post (check with a level) and hold it in the plumb position with wood braces. Then install a screw eye near the top of the gate post (right photo) and at the bottom of the next fence post. Install a turnbuckle cable between the two screw eyes and tighten it. Remove the braces and check the posts for plumb again. If any adjustment is required, tighten or loosen the turnbuckle.

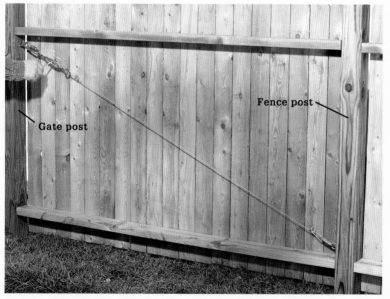

Gate post

Fence post

# Photo Credits

**California Redwood Association**
p. 56

**Todd Caverly**
p. 12 (top), 93 (sidebar top)

**Crandall & Crandall**
p. 149 (both)

**Tony Giammarino**
p. 6, 9 (lower left), 50–51, 93 (lower), 102

**iStock Photo**
p. 8 (lower), 12 (lower), 13 (top), 18

**Shelley Metcalf / shelley.metcalf@cox.net**
p. 7 (both), 9 (top) William Bocken Architecture and
    Interior Design, Paul Adams Landscape Design 619-
    260-1162, 9 (lower right), 11 (top left & lower), 64

**David Mortensen / featured on grassrootsmodern.com**
p. 11 (top right)

**Clive Nichols**
p. 162
Designer: Jane Mooney (bright blue wall and raised
    bed planted with herbaceous plants); Sculpture by
    John Brown.

**Jerry Pavia**
p. 4, 8 (top right), 10 (lower), 13 (lower left), 122 (left),
    134–135, 176

**Saxon Holt Photography / photobotanic.com**
p. 130 (top)

**TREX**
p. 82

**Walpole Woodworkers**
p. 125 (all)

# Resources

**Black & Decker**
Portable power tools and more
www.blackanddecker.com

**Cali Bamboo**
Bamboo fencing, flooring, and more
888.788.2254
www.calibamboo.com
Featured on p. 3, 10 (top left), 106–109

**California Redwood Association**
www.calredwood.com

**North American One-Call Referral System**
Call before you dig!
888.258.0808

**Red Wing Shoes Co.**
work shoes and boots shown throughout book
800 733 9464
www.redwingshoes.com

**Quikrete**
Cement and concrete products
800.282.5828
www.quikrete.com

**TREX**
Wood-alternative fencing products
800.289.8739
www.trex.com
Featured on p. 82–85.

# Conversion Charts

## Metric Equivalent

| Inches (in.) | $\frac{1}{64}$ | $\frac{1}{32}$ | $\frac{1}{25}$ | $\frac{1}{16}$ | $\frac{1}{8}$ | $\frac{1}{4}$ | $\frac{3}{8}$ | $\frac{2}{5}$ | $\frac{1}{2}$ | $\frac{5}{8}$ | $\frac{3}{4}$ | $\frac{7}{8}$ | 1 | 2 | 3 | 4 | 5 | 6 | 7 | 8 | 9 | 10 | 11 | 12 | 36 | 39.4 |
|---|---|---|---|---|---|---|---|---|---|---|---|---|---|---|---|---|---|---|---|---|---|---|---|---|---|---|
| Feet (ft.) | | | | | | | | | | | | | | | | | | | | | | | | 1 | 3 | 3$\frac{1}{12}$ |
| Yards (yd.) | | | | | | | | | | | | | | | | | | | | | | | | | 1 | 1$\frac{1}{12}$ |
| Millimeters (mm) | 0.40 | 0.79 | 1 | 1.59 | 3.18 | 6.35 | 9.53 | 10 | 12.7 | 15.9 | 19.1 | 22.2 | 25.4 | 50.8 | 76.2 | 101.6 | 127 | 152 | 178 | 203 | 229 | 254 | 279 | 305 | 914 | 1,000 |
| Centimeters (cm) | | | 0.95 | 1 | 1.27 | 1.59 | 1.91 | 2.22 | 2.54 | 5.08 | 7.62 | 10.16 | 12.7 | 15.2 | 17.8 | 20.3 | 22.9 | 25.4 | 27.9 | 30.5 | 91.4 | 100 | | | | |
| Meters (m) | | | | | | | | | | | | | | | | | | | | | | .30 | .91 | 1.00 | | |

## Converting Measurements

| To Convert: | To: | Multiply by: |
|---|---|---|
| Inches | Millimeters | 25.4 |
| Inches | Centimeters | 2.54 |
| Feet | Meters | 0.305 |
| Yards | Meters | 0.914 |
| Miles | Kilometers | 1.609 |
| Square inches | Square centimeters | 6.45 |
| Square feet | Square meters | 0.093 |
| Square yards | Square meters | 0.836 |
| Cubic inches | Cubic centimeters | 16.4 |
| Cubic feet | Cubic meters | 0.0283 |
| Cubic yards | Cubic meters | 0.765 |
| Pints (U.S.) | Liters | 0.473 (Imp. 0.568) |
| Quarts (U.S.) | Liters | 0.946 (Imp. 1.136) |
| Gallons (U.S.) | Liters | 3.785 (Imp. 4.546) |
| Ounces | Grams | 28.4 |
| Pounds | Kilograms | 0.454 |
| Tons | Metric tons | 0.907 |

| To Convert: | To: | Multiply by: |
|---|---|---|
| Millimeters | Inches | 0.039 |
| Centimeters | Inches | 0.394 |
| Meters | Feet | 3.28 |
| Meters | Yards | 1.09 |
| Kilometers | Miles | 0.621 |
| Square centimeters | Square inches | 0.155 |
| Square meters | Square feet | 10.8 |
| Square meters | Square yards | 1.2 |
| Cubic centimeters | Cubic inches | 0.061 |
| Cubic meters | Cubic feet | 35.3 |
| Cubic meters | Cubic yards | 1.31 |
| Liters | Pints (U.S.) | 2.114 (Imp. 1.76) |
| Liters | Quarts (U.S.) | 1.057 (Imp. 0.88) |
| Liters | Gallons (U.S.) | 0.264 (Imp. 0.22) |
| Grams | Ounces | 0.035 |
| Kilograms | Pounds | 2.2 |
| Metric tons | Tons | 1.1 |

## Converting Temperatures

Convert degrees Fahrenheit (F) to degrees Celsius (C) by following this simple formula: Subtract 32 from the Fahrenheit temperature reading. Then mulitply that number by $\frac{5}{9}$. For example, 77°F - 32 = 45. 45 × $\frac{5}{9}$ = 25°C.

To convert degrees Celsius to degrees Fahrenheit, multiply the Celsius temperature reading by $\frac{9}{5}$, then add 32. For example, 25°C × $\frac{9}{5}$ = 45. 45 + 32 = 77°F.

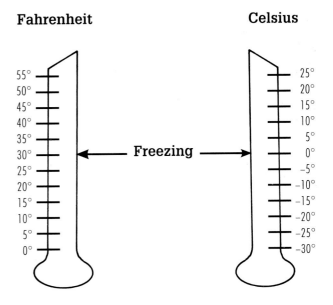

**Fahrenheit**      **Celsius**

Freezing

# Index

**A**
Aluminum gates, 125
Arbor gate idea, 9
Arched gates, building, 126–129

**B**
Bamboo
  about, 47
  chain link fences, covering with, 109
  idea, 10
  old fences, covering with, 109
  wood-frame fences, building,
    106–108
Bankers, 32
Batter, 30, 31
Batter boards, installing, 23
Blocks, interlocking concrete
  cutting, 171
  idea, 13
  retaining walls, building, 168–173
Board fences
  capped post &, building, 72
  modern privacy, building, 73
  post &, building, 68–70
  post &, building notched, 71
  stringer &, building, 52–55
Braces, securing on concrete
  slabs, 163
Brick & cedar fences, building,
  116–119
Bricks
  about, 47
  brick splitters, using, 37
  damaged, replacing, 194–195
  pillars, building, 116–119
  project planning, 34–35
  scoring & cutting, 36
Brick walls
  building, 154–157
  repairing, 192–195
Building codes
  footing requirements, 42
  planning and, 16

**C**
California-style chain link fences,
  installing, 99–100
Capped post & board fences,
  building, 72
Caps, flagstone, 33
Cedar & brick fences, building,
  116–119
Chain link
  about, 47
  weaving mesh together, 97
Chain link fences
  California-style installations,
    99–100

covering with bamboo, 109
  installing, 94–97
Chain link gates
  about, 125
  installing, 98
Columns
  brick, building, 116–119
  decorative, adding to wall, 141
Come-along tools, 196
Composite (wood)
  about, 47
  fences, building, 82–85
  fences, repairing, 187
Concrete
  colored, 184
  decorative forms, 183
  footings, 28, 42, 43–45
  fortifier, 192
  landscaping tools and, 48
  materials, 41
  mixing, 28, 40
  slabs, building, 165
  slabs, building walls on, 165
  slabs, securing braces on, 163
  tools, 49
  weight of cubic foot, 162
Concrete blocks
  about, 47
  cutting, 37
  mortarless walls, building with,
    158–161
  project planning, 34–35
Concrete blocks, interlocking
  cutting, 171
  idea, 13
  retaining walls, building, 168–173
Concrete blocks, landscape
  about, 47
  cutting, 140
  freestanding patio walls, building,
    136–141
  idea, 8
Concrete (poured) walls
  building, 162–167
  building retaining, 180–185
  idea, 10
Contoured installation, 18, 19, 20
Countertops, building outdoor,
  142–145
Curves
  dry stone walls, building with,
    137, 149
  interlocking block retaining walls,
    adding to, 173
  laying out, 25
Custom-built fences and slopes, 18

**D**
Decorative columns, adding to
  wall, 141
Design ideas, 6–13
Dog fences, invisible
  about, 110
  boundary layout options, 111
  installing, 112–115
Double-wythe constructions, 34, 154
Drainage
  improving, 189
  posts and, 29
Drawings, 17
Dry laying, 146
Dry stone walls
  building, 146–148
  building on slopes, 149
  building with curves, 137, 149
  repairing, 189

**E**
Elevation drawings, 17
Entry wall idea, 10

**F**
Face-mounted wood panel fences,
  building, 62–63
Facets, cutting in blocks, 140
Fencelines, laying out, 22–25
Fences
  bamboo, building wood-frame,
    106–108
  bamboo for covering old, 109
  board & stringer, building, 52–55
  brick & cedar, building, 116–119
  California-style chain link,
    installing, 99–100
  chain link, installing, 94–97
  chain link, installing privacy
    filling, 101
  invisible dog, installing, 110–115
  laying out lines for, 22–25
  materials for, 46–47
  modern post & board privacy,
    building, 73
  ornamental metal, building, 90–93
  picket, building, 64–67
  planning considerations, 5, 16–17
  post & board, building, 68–70
  post & board, building notched, 71
  post & board, building with
    capped rails, 72
  slopes, installing on, 18–21
  split rail, building, 74–78
  stone & rail, building, 120–121
  trellis, building, 102–105
  vinyl panel, building, 86–89

Virginia rail, building, 78–81
wood, painting, 201
wood composite, building, 82–85
wood panel, about, 56–57
wood panel, building, 58–61
wood panel, building face-
mounted, 62–63
Fences, repairing wood, 196
materials for, 197
painting, 201
posts and stringer, 198–199
straightening leaning, 200
tools for, 196–197
Flagstone, cutting, 33
Footings
building, 43
concrete, 28, 42, 43–45
frost, 35, 42
Freestanding patio block walls,
building, 136–141
Frost footings, 35, 42

**G**
Gates
aluminum, 125
arched, building, 126–129
chain link, about, 125
chain link, installing, 98
custom, 122
hardware, 47
idea, 9
iron, 125
metal, 125
perimeter-frame, building, 124
post spacing, 23
prefab, installing, 61
prefab, options, 125
for split rail fences, 78
trellis, building, 130–133
vinyl, installing, 89
Z-frame, building, 123
Gates, repairing
leaning posts, 203
tools for, 197
worn, 202–203

**H**
Hardware, about, 47
Head joints, 30
Hinge posts, 89

**I**
Interlocking concrete blocks
cutting, 171
idea, 13
retaining walls, building, 168–173
Invisible dog fences
about, 110
boundary layout options, 111
installing, 112–115
Iron gates, 125

**J**
Joints
repointing, 190
strong and weak, 30
tuck-pointing, 193

**K**
Kiln-dried lumber (KDAT), 46
Kitchen walls with countertops,
building, 142–145

**L**
Landscape blocks
about, 47
cutting, 140
freestanding patio walls, building,
136–141
idea, 8
Leaning gate posts, fixing, 203
Leaning wood fences,
straightening, 200
Line posts and gate posts, spacing, 23

**M**
Masonry tools, 49
Metal fences
about wrought iron, 93
building, 90–93
ideas, 6, 9, 10
repairs, 187
Metal gates, 125
Modern post & board privacy fences,
building, 73
Mortar
mixing & placing, 39
spills & oozes, cleaning, 153
stone walls, repairing with,
190–191
working with, 38
Mortared block walls idea, 10
Mortared stone walls
building, 150–153
repairing, 190–191
Mortar joints
repointing, 190
tuck-pointing, 193
Mortarless concrete block walls,
building, 158–161

**N**
Natural stone. *See* Entries beginning
with *stone*
Notched post & board fences,
building, 71

**O**
Ornamental metal
about, 46
cutting, 93
Ornamental metal fences
building, 90–93
idea, 10

Outdoor kitchen walls with
countertops, building, 142–145

**P**
Painting wood fences, 201
Patio blocks
about, 47
cutting, 140
freestanding patio walls, building,
136–141
idea, 8
Perimeter-frame gates, building, 124
Pet fences, invisible
about, 110
boundary layout options, 111
installing, 112–115
Picket fences
building, 64–67
idea, 11
painting, 201
repairing, 200
on slopes, 18
Pillars
brick, building, 116–119
decorative, adding to wall, 141
Planning
considerations, 5, 16
drawings/maps, 17
Plants
for trellis fences, 102
trellis fence ties, 105
Plot boundaries, 16
Post & board fences
building, 68–70
capped rail, building, 72
modern privacy, building, 73
notched, building, 71
Post infills, 89
Posts
contoured installations, 20
drainage and, 29
hinge, 89
leaning gate, fixing, 203
line and gate spacing, 23
ornamental metal fence spacing, 92
in picket fences, 65
rotted, repairing, 198–199
setting, 26–29
spacing, 17, 23
stepped installations, 20
Poured concrete walls
building, 162–167
idea, 10
retaining, building, 180–185
Prefabricated fence panels
post spacing, 17
on slopes, 19, 58
vinyl, building with, 86–89
wood, about, 56–57
wood, building, 58–61
wood, building face-mounted,
62–63

Prefabricated gate options, 125
Pressure-treated (PT) lumber, 46
Privacy fences
    bamboo, building wood-frame,
        106–108
    chain link, 101
    ideas, 8, 10
    modern post & board, building, 73
    vinyl panel, building, 86–89
    wood composite, building, 82–85

**R**
Racking a panel, 19
Rail fences, building
    split, 74–78
    stone &, 120–121
    Virginia, 78–81
Repointing joints, 190
Retaining walls
    building tips, 169
    ideas, 10, 13
    interlocking block, building,
        168–173
    positioning, 169
    poured concrete, building,
        180–185
    stone, building, 176–179
    timber, building, 174–175
Right angles, laying out, 24
Rotted posts & stringers, repairing,
    198–199

**S**
Shiners, 30
Single-wythe constructions, 34, 154
Site maps, 17
Slopes, managing
    with dry walls, 149
    options, 18–21
    panel fences for, 19, 58
    with post & board fences, 68
Split rail fences, building, 74–78
Splitters for bricks, using, 37
Stepped installations, 18–19, 21, 58
Stone & rail fences, building,
    120–121
Stone retaining walls, building,
    176–179
Stones
    about, 47
    cutting, 32–33
    damaged, replacing with
        mortar, 191
    popped, replacing, 188
    tools for, 49
Stone walls, building
    amount needed, estimating, 30
    dry, 146–148
    dry on slopes, 149
    dry with curves, 137, 149
    idea, 13

laying, 31
    mortared, 150–153
Stone walls, repairing
    dry, 189
    mortared, 190–191
Straightening leaning wood
    fences, 200
Stringers
    contoured installations, 18, 20
    fences of board &, building, 52–55
    rotted, repairing, 198–199
    stepped installations, 20
Stucco wall idea, 9

**T**
Tie-rods, 43
Tie stones, 30, 31
Timber retaining walls, building,
    174–175
Tools, 48–49, 196–197
Trellis fences, building, 102–105
Trellis gates, building, 130–133
Tuck-pointing mortar joints, 193
Turnbuckle cables, 196

**U**
Utility lines, 16

**V**
Vinyl fences
    about, 47
    gate installation, 89
    idea, 13
    panels, building with, 86–89
    repairs, 187
    on slopes, 18
Vinyl gates, 89, 125
Virginia rail fences, building, 78–81

**W**
Walls, building
    brick, 154–157
    on concrete slabs, 165
    dry stone, 146–148
    dry stone on slopes, 149
    dry stone with curves, 137, 149
    freestanding patio block, 136–141
    height of, 135
    interlocking block retaining,
        168–173
    materials for, 47
    mortared stone, 150–153
    mortarless concrete block,
        158–161
    outdoor kitchen with countertops,
        142–145
    planning considerations, 5, 16–17
    poured concrete, 162–167
    poured concrete retaining,
        180–185
    stone retaining, 176–179
    timber retaining, 174–175

Walls, repairing
    brick, 192–195
    dry stone, 189
    mortared stone, 190–191
Water runoff, 29, 189
Wood, about, 46
Wood composite
    about, 47
    fences, building, 82–85
    fences, repairing, 187
Wood fences, building
    board & stringer, 52–55
    brick & cedar, 116–119
    picket, 64–67
    planning considerations, 5, 16–17
    post & board, 68–70
    post & board notched, 71
    post & board privacy, 73
    post & board with capped rails, 72
    on slopes, 18–21
    split rail, 74–78
    stone & rail, 120–121
    trellis, 102–105
    Virginia rail, 78–81
Wood fences, repairing, 196
    materials for, 197
    painting, 201
    posts and stringer, 198–199
    straightening leaning, 200
    tools for, 196–197
Wood-frame bamboo fences,
    building, 106–108
Wood gates
    about custom, 122
    arched, building, 126–129
    perimeter-frame, building, 124
    trellis, building, 130–133
    Z-frame, building, 123
Wood panel fences
    about, 56–57
    building, 58–61
    building face-mounted, 62–63
    on slopes, 18
Wrought iron fencing
    about, 93
    idea, 6

**Z**
Z-frame gates, building, 123